Analysis of Peptides and Proteins by Electrophoretic Techniques

Analysis of Peptides and Proteins by Electrophoretic Techniques

Special Issue Editors

Angela R. Piergiovanni
José Manuel Herrero-Martínez

MDPI • Basel • Beijing • Wuhan • Barcelona • Belgrade

MDPI

Special Issue Editors

Angela R. Piergiovanni
Istituto di Bioscienze e Biorisorse
Italy

José Manuel Herrero-Martínez
University of Valencia
Spain

Editorial Office
MDPI
St. Alban-Anlage 66
4052 Basel, Switzerland

This is a reprint of articles from the Special Issue published online in the open access journal *Molecules* (ISSN 1420-3049) from 2018 to 2019 (available at: http://www.mdpi.com/journal/molecules/special_issues/electrophoretic_techniques)

For citation purposes, cite each article independently as indicated on the article page online and as indicated below:

LastName, A.A.; LastName, B.B.; LastName, C.C. Article Title. *Journal Name* **Year**, *Article Number, Page Range.*

ISBN 978-3-03921-227-9 (Pbk)
ISBN 978-3-03921-228-6 (PDF)

Contents

About the Special Issue Editors

Angela R. Piergiovanni obtained her degree in Chemistry at University of Bari (Italy), and successively attended the school of Chemical Science at University of Bari. She received a fellowship in association with the International Institute of Tropical Agriculture (I.I.T.A., Ibadan, Nigeria) at Istituto del Germoplasma of National Research Council (IG-CNR), Bari.

In 1990, she became Researcher at the Istituto di Genetica Vegetale, and is currently Senior Researcher at Istituto di Bioscienze e BioRisorse (IBBR-CNR) since her appointment in January 2010. She has participated in several projects and international working groups dealing with grain crops, is a member of several Italian Scientific Societies, and reviewer for more than 15 international scientific journals.

Her research areas include the study of genetic resources of herbaceous species relevant for Mediterranean agriculture, the analysis of protein fractions by using electrophoretic techniques (CE, SDS-PAGE and PAGE), and the evaluation of seed quality of Italian legume landraces.

She is author of more than 230 scientific contributions, where 45% are papers published in international peer-reviewed journals, books, or book chapters.

José Manuel Herrero-Martínez received his Ph.D. degree from University of Valencia (Spain) in 2000. He has worked as Assistant Professor (2001–2005) at Department of Analytical Chemistry (University of Barcelona, Spain) and as a Postdoctoral Researcher (2003–2004) at Department of Chemical Engineering (University of Amsterdam, The Netherlands). He is currently Full Professor at the Department of Analytical Chemistry, University of Valencia. He has published around 130 research articles and two book chapters. He has conducted extensive research in the area of capillary electromigration techniques and their application in food, biomedical, and industrial areas. His current research interests involve the synthesis and characterization of novel stationary phases for separation techniques and sample treatment based on polymeric materials and composites.

Preface to "Analysis of Peptides and Proteins by Electrophoretic Techniques"

The characterization of complex matrices containing peptides and proteins is a relevant issue in the research of life and biological sciences. To understand the key role of these macromolecules in the structure and function of cells belonging to animal or plant tissues, as well as in nutritional, physicochemical, and sensorial food traits, the study of their expression levels, post-translational modifications, and specific interactions is necessary. The first step of these investigations consists in the extraction of proteins and peptides from real matrices using appropriate methodologies. Regardless of the starting tissue and the effectiveness of the used extraction method, mixtures of proteins or peptides with similar chemicophysical properties provide a starting sample for subsequent detailed analysis. In order to characterize each component of these mixtures, powerful separation techniques are required. In addition to chromatographic methods, electrophoretic techniques are known to represent a broad and powerful family of methodologies able to separate, visualize, and quantify single proteins or peptides. A large part of these techniques is automated, allowing for processing of a high number of samples. Moreover, in the last decade, the development of microdevices has reduced sample consumption and waste production while use of high-sensitivity detectors, such as mass spectrometry (MS) or laser-induced fluorescence (LIF), have significantly improved with regards to separation efficiency and detection limits. All of these advancements have enlarged the field of application for electrophoretic techniques.

This Special Issue of Molecules, entitled "Analysis of Peptides and Proteins by Electrophoretic Techniques", covers some of the recent and relevant advancements with regard to this subject matter. This issue includes three research papers describing the use of capillary electrophoresis (CE) protocols and slab gels to separate and characterize macromolecules present in biological matrices of clinical interest. Toxicology is the field of investigation in the fourth paper, which characterizes the venom proteome of an African spitting cobra species using 2-D electrophoresis and MALDI ToF/ToF (matrix-assisted laser desorption/ionization time of flight) mass spectrometry techniques.

The two reviews included in this issue present the state of the art regarding the use of CE methodologies in specific fields of application. The first reports on the expansion of immune and enzyme assay portfolios obtained using CE-LIF while the second addresses progress on the biology of seed storage proteins and their application in breeding using two-dimensional electrophoresis (2-DE)-based maps.

Angela R. Piergiovanni, José Manuel Herrero-Martínez

Special Issue Editors

molecules

MDPI

Article

Carbon Dot-Mediated Capillary Electrophoresis Separations of Metallated and Demetallated Forms of Transferrin Protein

Leona R. Sirkisoon [1], Honest C. Makamba [2], Shingo Saito [3] and Christa L. Colyer [1,*]

[1] Department of Chemistry, Wake Forest University, Winston-Salem, NC 27109, USA; sirklr12@wfu.edu
[2] Razzberry Inc., 5 Science Park, Unit 2E9, New Haven, CT 06511, USA; honest737@gmail.com
[3] Graduate School of Science and Engineering, Saitama University, Saitama 338-8570, Japan; shingo@mail.saitama-u.ac.jp
* Correspondence: colyercl@wfu.edu; Tel.: +81-336-758-4936

Received: 2 May 2019; Accepted: 16 May 2019; Published: 18 May 2019

Abstract: Carbon dots (CDs) are fluorescent nanomaterials used extensively in bioimaging, biosensing and biomedicine. This is due in large part to their biocompatibility, photostability, lower toxicity, and lower cost, compared to inorganic quantum dots or organic dyes. However, little is known about the utility of CDs as separation adjuvants in capillary electrophoresis (CE) separations. CDs were synthesized in-house according to a 'bottom-up' method from citric acid or other simple carbon precursors. To demonstrate the applicability of CDs as separation adjuvants, mixtures of holo- (metallated) and apo- (demetallated) forms of transferrin (Tf, an iron transport protein) were analyzed. In the absence of CDs, the proteins were not resolved by a simple CE method; however, upon addition of CDs to the separation buffer, multiple forms of Tf were resolved indicating that CDs are valuable tools to facilitate the separation of analytes by CE. CE parameters including sample preparation, buffer identity, ionic strength, pH, capillary inside diameter, and temperature were optimized. The results suggest that dots synthesized from citric acid provide the best resolution of various different forms of Tf and that CDs are versatile and promising tools to improve current electrophoretic separation methods, especially for metalloprotein analysis.

Keywords: carbon dots; capillary electrophoresis; transferrin; metalloproteins; fluorescence

1. Introduction

Carbon dots (CDs) are a unique type of fluorescent nanomaterial consisting of a graphene core decorated with oxygenated functional groups on the surface [1–5]. They are structures comprising of one to a few layers of graphene sheets smaller than 10 nm in diameter. The distinctive photoluminescence of CDs is attributed to the sp^2 hybridized carbon atoms and the quantum confinement and edge effects resulting from the small size of these carbon-based materials [6]. For example, typical CDs synthesized from citric acid exhibit an emission maximum at 460 nm, independent of excitation wavelength from 300–420 nm, with carboxylic acid and hydroxyl functional groups on the surface [1,6]. CDs interact with potential analytes through hydrophobic, π-π stacking, hydrogen bonding, cation-π, and electrostatic interactions. The dispersibility of CDs in aqueous solutions is due to the hydroxyl and carbonyl functional groups on their surface, which can be easily altered to render the materials hydrophobic or amphiphilic [2]. CDs exhibit characteristic chemical and physical properties such as biocompatibility, photostability, and low toxicity, and they have the added advantages of simple and low cost synthesis methods. These features have triggered interest in the use of CDs as alternative fluorescence probes in place of organic dyes and inorganic nanoparticles [1,3,5–7]. Many recent applications involving CDs capitalize on their fluorescent properties for bioimaging [8–11], biomedicine [12], and biosensing [13–15] to aid in the diagnosis and treatments of diseases, defects, and cancers [1]. However, little is known about the utility of CDs as separation adjuvants in capillary electrophoresis (CE) [3] in comparison to other nanomaterials such as silica nanoparticles [16,17], carbon nanotubes [18], graphene nanoparticles [19], single-walled carbon nanotubes [20], and gold nanoparticles [21–23], which have all been reported to enhance CE separations.

CE is a high resolution separation technique that separates analytes based on differential migration rates of charged species in an electric field [24,25]. Advantages of CE include relatively fast analysis times, high efficiency separations, and small sample volumes [3,26]. Further selectivity may be achieved in CE by employing pseudo-stationary phases (solution-based additives present in the separation buffer, which effect the separation of analytes based on their differential associations). The use of pseudo-stationary phases rather than true stationary phases in CE-based methods reduces problems with irreproducibility between capillaries and furthermore, it is simpler than introducing selectivity via the more time-consuming process of immobilization of nanomaterials to form inner capillary wall coatings [27,28]. While surfactants are among the most commonly encountered buffer additives in CE, the use of soluble nanomaterials as buffer additives (or "separation adjuvants") provides another option for CE method development. For example, Sun and colleagues [3] successfully employed CDs as additives for the separation of cinnamic acid and its derivatives by CE coupled with UV detection and observed increased resolution between cinnamic acid and its derivatives, concluding that CDs are a promising separation material for analytical methods. While carbon nanotubes have be used to assist in protein separations by CE [29], there are no published reports of CDs being used in this capacity and thus, the potential for new developments in this area remains great. Based on these (limited) precedents, we have sought to advance our understanding not only of the versatility and utility of CDs as CE separation adjuvants but also, of metallated protein separations by CE.

In particular, this work focuses on the separation of transferrin (Tf) protein. Tf is a globular, iron transport glycoprotein (comprised of 679 amino acid residues with a molecular weight of 80 kDa). It has two lobes (the N and C lobes) with a high affinity Fe^{3+} binding domain in each [30,31]. When iron is bound to both lobes in Tf (constituting the fully metallated or "holo-" form of the protein), the protein adopts a structural conformation that is more closed (folded) than that of the demetallated ("apo-") Tf protein. There are four possible conformations of Tf, depending on the number and position of bound Fe^{3+} ions: (i) holo-Tf (fully metallated), (ii) single Fe^{3+} bound only to the C-lobe or (iii) only to the N-lobe (partially metallated), and (iv) apo-Tf (demetallated). The Tf receptor is overexpressed on proliferating cancer cells, but not normal cells; therefore, Tf is a promising carrier protein for targeted drug delivery and therapy for cancerous cells [31–39]. The ability to separate the different conformations of Tf (fully metallated, partially metallated, and demetallated), is important because potential drug molecules may have different affinities for the different conformations of Tf. However, a major challenge in separating apo- and holo-Tf by CE is the fact that bound metal ions exert only subtle changes in overall protein mass and charge [40]. This challenge may be met by the use of pseudostationary phases or buffer additives, as demonstrated previously by Nowak and colleagues [26,40], who developed and optimized a CE method for the separation of different forms of Tf using micellar electrokinetic chromatography. Their work, employed sodium dodecyl sulfate and 20% methanol as separation buffer additives, leading to the resolution of apo-Tf, holo-Tf, two partially metallated forms of Tf, lactoferrin, and human serum albumin proteins.

Just as Nowak's use of surfactants in CE was able to afford greater resolution of metallated and demetallated protein forms, we hypothesized that the use of CDs in CE should likewise afford the necessary selectivity for Tf separations. To this end, CDs were synthesized in-house by pyrolysis of citric acid and other organic precursors. Fluorescence studies were performed to assess the interaction between CDs and apo- and holo-Tf. A significant quenching was observed for the mixture of CDs with holo-Tf and no change in fluorescence signal was observed for CDs with apo-Tf, suggesting that the extent of protein metallation has an impact on protein interaction with CDs. A mixture of holo- and apo-Tf was analyzed by a simple CE method. In the absence of CDs, the proteins were not resolved; however, upon addition of CDs to the separation buffer, multiple forms of Tf were resolved. Sample preparation, buffer identity, ionic strength, pH, capillary inside diameter, and temperature were optimized. The results indicate that dots synthesized from citric acid provide the best resolution between the different metallated forms of Tf. Results from this work indicate that CDs are inexpensive, stable, and convenient buffer additives able to improve current electrophoretic separations of metalloproteins, with implications for greater selectivity in the CE separations of other classes of analyte.

2. Results and Discussion

2.1. Probing Interactions Between CDs and Tf by Fluorimetry

The interactions between CDs and metallated versus demetallated forms of Tf were assessed by fluorimetry. CDs used in these studies were synthesized by oven pyrolysis of dry citric acid reagent followed by suspension of the resulting CDs in aqueous solution. Fluorescence emission of the CDs alone was measured, followed by emission of the CDs upon addition of increasing amounts of apo-Tf or holo-Tf, as seen in Figure 1. No significant change (11.2% quenching) in fluorescence emission (at 460 nm) was observed for a 35 µg/mL CD sample as the concentration of apo-Tf was increased from 0 to 100 µM (Figure 1A). However, the fluorescence signal was quenched by as much as 47.6% upon the addition of up to 100 µM holo-Tf to the same CD sample (see Figure 1B). The intensities represented in Figure 1C were determined at the wavelength of maximum fluorescence emission (460 nm) after having corrected for native Tf fluorescence at each concentration (as shown in Figure S1A,B) and applying a five-point boxcar smoothing. The extent of change in fluorescence of CDs as a function of Tf protein concentration is represented by the slopes of the response curves in Figure 1C. The slope for apo-Tf is −0.0112 RFU/µM indicating very little to no change in fluorescence of the CDs. However, the slope for holo-Tf is −0.048 RFU/µM, revealing a direct proportionality between the extent of fluorescence quenching of the CDs signal and the concentration of holo-Tf. In work by Bhattacharya and colleagues [39] a similar effect was characterized as static quenching via their steady-state and time-resolved photoluminescence measurements at pH 7.4. Based on estimated thermodynamic parameters of the CD-Tf association determined from quenching measurements performed at various temperatures, they concluded that the observed quenching was a result of the electrostatic interaction between CDs and the Fe^{3+} ions associated with holo-Tf, not the amino acid residues. Furthermore, Zhu and coworkers [41] showed that the presence of Fe^{3+} ions in bulk solution quenched the intrinsic fluorescence of CDs. Therefore, we believe the differential effect of apo- versus holo-Tf on the fluorescence of CDs in our experiments is most likely a result of the paramagnetic property of the Fe^{3+} ions of the holo-Tf impacting the quantum yield. However, such an effect does not preclude the possibility of different metallated protein states interacting to different extents with the CDs (and we explore this possibility in more detail in the capillary electrophoresis studies discussed in Section 2.2).

Additionally, the experiment was repeated using CDs synthesized in the autoclave and suspended in aqueous solution. A similar trend was observed with these CDs: little to no change in fluorescence emission of the CDs upon increasing the concentration of apo-Tf (Figure S-1D), and decreased fluorescence emission upon increasing the concentration of holo-Tf (Figure S-1E). The conformation of demetallated apo-Tf is such that it has two tryrosine, one aspartate, and one histidine residue exposed [39]. While it seems plausible that these exposed residues could interact with the CDs (via hydrophobic, π-π stacking, H-bonding, or electrostatic interactions), the relative lack of change in fluorescence emission of apo-Tf with CDs could not provide evidence for any such interactions under the solution conditions employed here. However, the observed fluorescence quenching of CDs with holo-Tf indicates that the bound Fe^{3+} in the metallated form of the protein experiences electrostatic interactions with the hydroxyl and carboxylic acid groups on the surface of the CDs, resulting in a non-emissive ground state complex [39]. Thus, even though CDs may interact (to a different extent) with demetallated and metallated forms of Tf, this could not be confirmed by fluorescence studies alone.

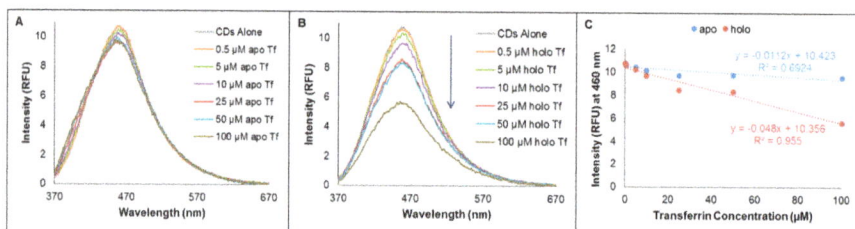

Figure 1. Fluorescence emission spectra for 35 µg/mL samples of oven-synthesized citric acid CDs, with increasing concentrations (from 0.5 µM to 100 µM) of added apo-Tf (**A**) and holo-Tf (**B**). Fluorescence response in terms of intensity at the wavelength of maximum emission (460 nm) as a function of Tf concentration, for apo- and holo-Tf are shown in (**C**). All samples were prepared to the concentrations indicated in the Figures using 50 mM tris-200 mM tricine (pH 7.4) buffer as diluent. The data were corrected for the respective native Tf fluorescence at each concentration. The excitation wavelength was 360 nm and the emission scan range was 365–700 nm.

2.2. CE Method Development and Optimization for the Separation of Apo-Tf and Holo-Tf

2.2.1. Studying the Effects of Sample Preparation: Diluent and Sample Additives

Given the differential interactions of CDs with metallated versus demetallated forms of Tf, as evidenced by differences in fluorescence quenching (Section 2.1), we surmised that CDs might be useful in the separation of these protein forms. Samples of apo-Tf, holo-Tf, and mixtures of apo- and holo-Tf were first prepared in aqueous solution alone and then subjected to analysis by CE with UV absorbance detection, employing a 50 mM tris-200 mM tricine (pH 7.4) separation buffer. Typical electropherograms resulting from these protein samples prepared in aqueous solution–with no CDs–are shown in Figure 2A. Subsequently, the water-based Tf samples and the separation buffer were prepared with added CDs (such that the final concentration of dots was 35 µg/mL in all cases), and the resulting electropherograms are shown in Figure 2B. The CDs used for these CE experiments were synthesized from citric acid by oven pyrolysis, followed by suspension in 50 mM NaOH and dialysis against ultrapure water for eight hours prior to use, unless otherwise stated.

The blue traces (i) in Figure 2A,B represent apo-Tf samples without and with added CDs, respectively. While there was no significant change in the observed migration time of the apo-Tf peak as a result of adding CDs to the sample (and separation buffer), there was a marked change (50.6%) in the (negative) electrophoretic mobility μ_{ep} of apo-Tf (from -0.00239 cm^2 V^{-1} s^{-1} in the absence of CDs to -0.00360 cm^2 V^{-1} s^{-1} in the presence of CDs). This change in (negative) electrophoretic mobility of apo-Tf was accompanied by a 34.0% increase in peak height and a 44.4% increase in peak area. The increase in (negative) electrophoretic mobility may provide evidence of the association of apo-Tf with CDs to produce a larger complex with greater net negative charge. Such a complex with greater negative electrophoretic mobility would move counter to the direction of electroosmotic flow, and so might be expected to appear at a longer migration time in the resulting electropherogram. However, based on the position of the small negative marker peak in Figure 2A,B, the electroosmotic mobility was found to increase by 5.4% (from 0.0205 cm^2 V^{-1} s^{-1} to 0.0216 cm^2 V^{-1} s^{-1}) upon the addition of CDs to the buffer system. In this particular case, the combination of the increased electroosmotic mobility and the decreased (i.e., increased negative) electrophoretic mobility resulted in very little change in the net mobility of apo-Tf (with and without added CDs) and thus the migration time of the apo-Tf peak appeared virtually unchanged. The increase in apo-Tf peak height and area in the system containing CDs may provide further evidence of the formation of apo-Tf-CD complexes, since such complexes may demonstrate some variation in size and enhanced absorbance relative to free apo-Tf.

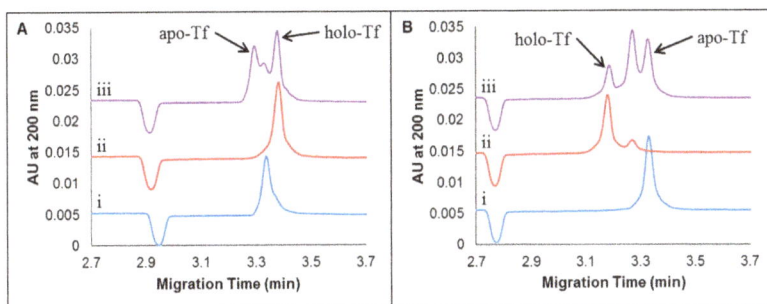

Figure 2. Effects of oven CDs as additives for samples of apo-Tf, holo-Tf and mixtures of apo- and holo-Tf (25 μM each) without CDs (**A**) and with CDs (**B**) for 25 μM apo-Tf (i), 25 μM holo-Tf (ii), and a mixture of apo- and holo-Tf (iii). Electropherograms are vertically offset for clarity. A volume of 1.25 nL (5.2 sec at 1.3 psi) was injected and 20 kV was applied. The separation occurred on a Beckman Coulter P/ACE MDQ System coupled with a UV detector at 15 °C on a 25 μm i.d. capillary with an effective length of 30 cm and a total length of 40 cm.

The red traces (ii) in Figure 2A,B represent holo-Tf samples without and with added CDs, respectively. A 6.0% decrease in migration time of holo-Tf (from 3.38 min to 3.18 min) was observed upon the addition of CDs. This reduced migration time is due to an increase in net mobility, and recall that net mobility is given by the sum of electroosmotic and electrophoretic mobilities. In the case of holo-Tf, it appears that the impact of added CDs on the electroosmotic flow (recall, a 5.9% increase in electroosmotic mobility was observed) was greater than the impact of added CDs on the electrophoretic mobility of the protein. The electrophoretic mobility of holo-Tf was found to be -0.00281 cm^2 V^{-1} s^{-1} in the absence of CDs and -0.00278 cm^2 V^{-1} s^{-1} in the presence of CDs, which represents just a 1.1% decrease (in the negative electrophoretic mobility, which is effectively the same as a 1.1% increase in μ_{ep} towards the cathode). This change is small in comparison to the 50.6% change in electrophoretic mobility observed for apo-Tf, which might suggest that the demetallated form of the protein has a greater affinity for (or forms more stable, long-lived complexes with) CDs compared to the metallated form of the protein. Thus, in the case of holo-Tf, the relatively small change in electrophoretic mobility is overshadowed by a greater change in electroosmotic flow upon the addition of CDs to the sample and separation buffers, which translates into a greater net mobility and shorter migration time.

The peak height of the primary holo-Tf peak decreased 19.1% and the area increased by 15.5% upon the addition of CDs (Figure 2A(ii) vs. Figure 2B(ii)). The decrease in peak height and increase in peak area is attributed to the loss of Fe^{3+} ions by holo-Tf [40] while the appearance of a new, smaller peak at 3.25 min (see Figure 2B(ii)) is attributed to a partially metallated form of Tf, which may associate with CDs in the separation buffer to a different extent than does the fully metallated form of Tf from which it originates. This appearance of an additional peak induced by the addition of CDs to the holo-Tf sample, taken together with changes in migration times or net mobilities, supports the idea of differential interactions between CDs with various different metallated forms of Tf.

Whereas samples of individual Tf proteins in the absence of CDs gave rise to single peaks (Figure 2A(i and ii)), a sample mixture containing 25 μM each of apo- and holo-Tf in water (also in the absence of CDs) gave rise to an unresolved cluster of three peaks by CE (Figure 2A(iii)). In the protein mixture, there is presumably an opportunity for exchange of Fe^{3+} ions between protein forms, resulting in unresolved metallated, demetallated, and partially metallated Tf proteins. Upon the addition of CDs, the cluster of three peaks was more clearly resolved in the electropherogram for the mixed-protein sample (Figure 2B(iii)). Interestingly, the combined area of the mixture increased 22.6%, and the migration order of apo-Tf and holo-Tf was reversed in the electropherogram of the protein mixture upon the addition of CDs to the sample and separation buffer. Whereas holo-Tf migrated last in the sample containing a mixture of proteins in the absence of CDs, it migrated first in the sample containing CDs. As discussed previously, this change in the proteins' net mobilities, brought about by the addition of CDs to the buffer system, may be attributed to the combined effects of a change in electroosmotic mobility and a change in electrophoretic mobility due to associations between CDs and Tf proteins. The overall impact was improved resolution of the protein mixture.

To further ascertain the importance of sample composition on CE resolution, Tf samples were prepared using the separation buffer (50 mM tris-200 mM tricine, pH 7.4) as a diluent rather than using pure water, without or with added CDs (35 µg/mL). Representative electropherograms are shown in Figure S2-A,B, respectively. Additionally, Tf samples were prepared in the buffer of 25 mM tris-100 mM tricine (pH 7.4). Representative electropherograms for these Tf samples without added CDs and with 17.5 µg/mL added CDs are shown in Figure S2-C,D, respectively. No significant improvement (nor deterioration) in separation efficiency was afforded by the changes sample buffer concentrations studied.

A comparison of Figure 2 and Figure S-2 leads us to conclude that an enhancement of the CE separation of apo- and holo-Tf is achieved in the presence of CDs regardless of sample composition. That is, preparations of Tf samples in water, separation buffer, and diluted separation buffer all resulted in similar electropherograms. The electropherograms for mixed samples containing both apo-Tf and holo-Tf protein standards revealed the appearance of a third peak, which was better resolved upon the addition of CDs to the sample and separation buffer. The appearance of this third peak upon mixing apo-Tf and holo-Tf together may indicate a partial exchange of Fe^{3+} from the holo-Tf to apo-Tf when mixed. Intraconversion between metallated and demetallated forms of Tf has been documented elsewhere [40]. In all cases, resolution improved upon the addition of CDs. In Figure 2, for example, the peak attributed to a partially metallated Tf species is better resolved from the apo-Tf peak ($R_s = 0.5$ without CDs and $R_s = 1.1$ with CDs) and it is also better resolved from the holo-Tf peak ($R_s = 0.8$ without CDs and $R_s = 1.5$ with CDs) in mixed protein samples. This suggests that the CDs interact differentially with each form of Tf, presumably due to differing contributions from hydrophobic, π-π stacking, H-bonding, or electrostatic interactions in the absence and presence of metal in various folded states of the proteins. Regardless of the sample preparation (that is, protein in water, 25 mM tris-100 mM tricine, or 50 mM tris-200 mM tricine), apo-Tf migrated first and holo-Tf last in the absence of CDs; however, in the presence of CDs, holo-Tf migrated first and apo-Tf last. Furthermore, since the effect of sample buffer ions on the resolution of a mixture of Tf protein forms was nominal relative to the effect of added CDs, method development is not constrained to a single sample preparation, giving the analyst greater flexibility when optimizing metalloprotein separations by this CD-enhanced CE method. Based on simplicity, ultrapure water with added CDs was chosen for Tf sample preparations in subsequent studies.

Whereas CDs were introduced simultaneously to both the sample preparation and the separation buffer to improve the separation of mixtures of apo-Tf and holo-Tf, as described above, the impact of CDs as separation adjuvants for on-column use only (CDs only in the separation buffer) and pre-column use only (CDs only in the sample preparation) was also explored. Pre-column use of CDs (as additives to the sample preparation only) did not result in a significant improvement in resolution of Tf protein forms (Figure S-3ii) relative to the use of no added CDs (Figure S-3i). However, CDs added to the separation buffer alone led to improved resolution of a mixture of Tf proteins relative to separations conducted without added CDs, as seen in Figures S-3-iii and S-3-iv relative to S-3-i. The resolution achieved with CDs in the separation buffer alone was still not as good as the resolution achieved with CDs in both the sample buffer and the separation buffer (Figure S-3-iv, and previously, Figure 2B-iii), and so CDs were employed as additives to both sample and separation buffers in all following CE experiments.

The effects of changes to CD composition on the resolution of the three Tf peaks observed for a mixture of apo- and holo-Tf with CDs were tested. Carbon dot composition was altered by replacing citric acid as the organic precursor with ascorbic acid, gluconic acid, N-acetylneuraminic acid, or glucose. Figure S-4 shows representative electropherograms employing these altered CDs

(35 µg/mL) as adjuvants in the separation of mixtures of apo- and holo-Tf with UV detection at 200 nm (Figure S-4A) and with LIF detection using a 375 nm laser and a 400 nm long pass filter (Figure S-4B). Altering dot composition by using different precursors resulted in some CDs with the potential to improve the resolution of the individual components in a mixture of apo- and holo-Tf with further optimization and others that did not affect the mobility at all as observed by UV detection and a single peak or broad hump from LIF detection. Although all of the chosen precursors result in CDs decorated with hydroxyl and carboxylic acid functional groups, the differences in their interaction with apo- and holo-Tf could be due to the ratio of carboxylic acid to hydroxyl functional groups on the surface, or differences in the carbon dot core, which may result from the arrangement of each precursor molecule as the carbon dot core was built, leading to the differences in the intrinsic fluorescence of the CDs from each precursor. Overall, CDs prepared from citric acid yielded best resolution between the metallated, partially metallated, and demetallated forms of Tf.

Incubation time studies ranging from 2–197 min (time elapsed between preparation of Tf protein samples with added CDs and their analysis by CE) revealed no correlation between sample incubation time and peak area or migration time (data not shown). Thus, CDs can be employed as separation adjuvants in CE studies without imposing any additional restrictions on method or time of sample preparation. This lends further credence to the utility of CDs as CE separation adjuvants.

2.2.2. Effect of Concentration of Added CDs

CE experiments were conducted with different concentrations of CDs added to the sample preparations and separation buffer in order to determine the optimal concentration to enhance the separation of a mixture of apo- and holo-Tf. The concentrations of CDs tested were 2, 5, 7, 10, 25, 35, 50, 75, 100, 250 and 500 µg/mL. A subset of these representative electropherograms are shown in Figure 3 (with the full concentration range shown in Figure S-5). At CD concentrations of 2–7 µg/mL, only two peaks were observed for a sample mixture containing 25 µM each of apo-Tf and holo-Tf (Figure 3-i). The addition of 10 µg/mL CDs gave rise to a broad signal with three unresolved components (Figure 3-ii). Upon the addition of anywhere from 25–500 µg/mL of CDs to the sample and separation buffer, three nearly resolved peaks were observed (Figure 3iii–v). It should be noted that sample compositions in Figure 3 differ from the optimal sample conditions shown in Figure 2 (optimal), due to the sequencing of experiments conducted. The samples in Figure 3 were prepared in 25 mM tris-100 mM tricine buffer (pH 7.4), with a concentration of CDs in the sample equal to half of the concentration in the separation buffer.

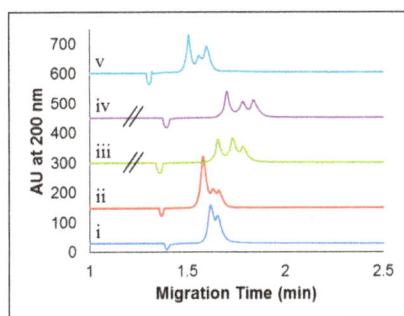

Figure 3. Abbreviated range of CDs concentrations tested with a mixture of apo- and holo-Tf (25 µM each). Concentrations of CDs shown are (i) 2 µg/mL, (ii) 10 µg/mL, (iii) 35 µg/mL, (iv) 100 µg/mL, and (v) 500 µg/mL. Electropherograms are vertically offset for clarity. A volume of 5 nL (2.1 s at 45 mbar) was injected and 20 kV was applied. The separation occurred on an Agilent G1600A CE coupled with a DAD UV/Vis Detector at 25 °C on a 50 µm i.d. capillary with an effective length of 24 cm and a total length of 32.5 cm.

Based on the results in Figure 3 (and Figure S-5), we determined the optimal concentration of CDs to be 35 μg/mL for the CE separation of the sample mixture of apo- and holo-Tf. While 25–500 μg/mL CDs also permitted the partial resolution of three peaks (attributed to apo-Tf, holo-Tf, and partially metallated Tf in the sample), the use of 35 μg/mL CDs was chosen as a conservative value to afford the necessary separation while also accommodating any synthetic variations from different batches of CDs, or effects due to post synthesis clean-up, and to prevent a high baseline from the absorbance of the CDs should they have been used at higher concentrations.

2.2.3. Separation Buffer Composition: Background Electrolyte, pH, and Concentration Effects

A variety of different background electrolytes were tested as separation buffers in order to determine their effects on separation efficiency for sample mixtures containing apo- and holo-Tf. These included buffers composed of phosphate, tris-tricine, tris-glycine, and tris-HCl, all at pH 7.4. Representative electropherograms obtained using each of these separation buffers for the analysis of a sample mixture containing 25 μM each of apo-Tf and holo-Tf with 35 μg/mL CDs are shown in Figure 4i–v. Tris-tricine and tris-HCl separation buffers at pH 7.4 (Figure 4-ii and Figure 4-v, respectively) afforded the best resolution (with three nearly resolved peaks representing apo-Tf, holo-Tf, and partially metallated Tf), albeit with the longest migration times relative to the other separation buffers tested. Calculated resolution values are summarized in Table S-2. The remaining separation buffers at pH 7.4 (10 mM phosphate, Figure 4-i; 50 mM tris-200 mM glycine, Figure 4-iii; 50 mM tris-500 mM glycine, Figure 4-iv) yielded faster eluting, unresolved peaks and so were not preferred above the tris-tricine and tris-HCl buffers.

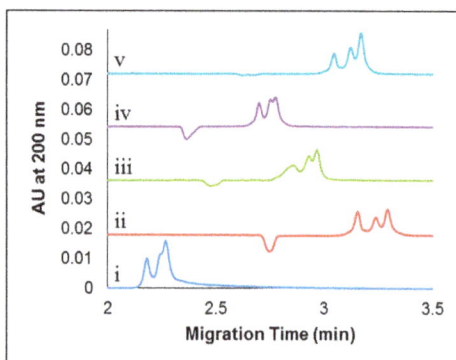

Figure 4. Separation buffer composition study for mixtures of apo- and holo-Tf (25 μM each) with CDs at pH 7.4 in different separation buffers. The separation buffers used were 10 mM phosphate (i), 50 mM tris-200 mM tricine (ii), 50 mM tris-200 mM glycine (iii), 50 mm tris-500 mM glycine (iv), and 50 mM tris-HCl (v). Electropherograms are vertically offset for clarity. A volume of 1.25 nL (5.2 s at 1.3 psi) was injected and 20 kV was applied. The separation occurred on a Beckman Coulter P/ACE MDQ System coupled with a UV detector at 25 °C on a 25 μm i.d. capillary with an effective length of 30 cm and a total length of 40 cm.

Subsequently, the effect of separation buffer pH on the resolution of a mixture of apo- and holo-Tf with CDs was evaluated with phosphate and tris-tricine separation buffers at pH 4.4, 7.4, and 10.4. No signal was observed in electropherograms recorded at pH 4.4 for both tris-tricine and phosphate buffers. At pH 7.4 the tris-tricine buffer gave rise to three peaks while the phosphate buffer gave rise to only two peaks (Figure S-6-i and S-6-ii, respectively), while at pH 10.4 only one peak was observed for both tris-tricine and phosphate buffers (Figure S-6-iii and S-6-iv, respectively). The lack of resolution afforded by pH 4.4 and 10.4 separation buffers and by phosphate buffer relative to tris-tricine buffer at all pHs tested, led us to conclude that the tris-tricine buffer at pH 7.4 (with CDs) was optimal for the resolution of a sample mixture containing apo-Tf and holo-Tf (also with CDs). However, the optimal concentration for the tris-tricine buffer remained to be determined, and so a concentration study was undertaken, as described presently.

With a fixed concentration of 50.0 mM tris, we varied the concentration of tricine from 100.0–300.0 mM to create a series of separation buffers (each adjusted to pH 7.4, if necessary) in order to determine the optimum buffer concentration for this work. Representative electropherograms recorded for a Tf mixture sample using the various concentrations of tris-tricine separation buffer (at pH 7.4, both with and without CDs) are shown in Figure S-7. At or above tricine concentrations of 200 mM we observed a significant increase in migration time for Tf; however, the increased resolution afforded by these higher concentration buffers, especially in the range of 200–250 mM tricine relative to 100–175 mM tricine, suggested that 200 mM tricine was optimal.

Using this, we subsequently varied the tris concentration from 25.0–100.0 mM in the separation buffer while maintaining a pH of 7.4 to complete the buffer optimization process. Representative electropherograms for a sample mixture of apo- and holo-Tf revealed three peaks for 200 mM tricine separation buffers containing CDs with either 25 mM or 50 mM tris (Figure S-8-i and S-8-ii, respectively), but only two peaks were resolved with the higher concentrations of tris in the separation buffer (75 mM, Figure S-8(iii); 100 mM, Figure S-8(iv)). At all concentrations, we again verified that the presence of CDs (in the tris-tricine buffer and the Tf sample) was essential to achieving resolution of the various metallated forms of the protein (Figure S-8). Hence, 50 mM tris-200 mM tricine (pH 7.4) containing 35 μg/mL CDs was chosen as the optimal separation buffer for this method.

2.2.4. Optimizing Capillary Inside Diameter and Temperature

In addition to optimizing the separation buffer and sample preparation including CDs, we likewise studied the effects of capillary inside diameter and temperature on the resolution of a mixture of apo- and holo-Tf in order to optimize the overall separation method. Separations were conducted using 20, 25, and 50 μm i.d. capillaries (as shown in Figure S-9A, S-9B, and S-9C, respectively), each thermostated at 15, 25, or 30 °C, with the optimized buffer and sample conditions determined herein. Interestingly, variation in capillary inside diameter and temperature within the ranges conducted in this study did not have a marked impact on separation efficiencies. As expected, increased temperatures led to decreased migration times (due to reduced buffer viscosities), and the largest capillary (50 μm i.d.) produced broader, less resolved signals with greater absolute absorbances (due to greater sample loading). Based on these results, the 25 μm i.d. capillary operated at 15°C (Figure S-9Bi) was chosen as optimal for this method.

Thus, the final optimized CE method, designed to afford the greatest resolution of a sample mixture containing various metallated forms of Tf protein, employs a 25 μm i.d. capillary at 15 °C with a 50 mM tris-200 mM tricine separation buffer (pH 7.4) containing 35 μg/mL CDs, and samples prepared or sample buffer with 35 μg/mL CDs added.

3. Materials and Methods

3.1. Reagents

Citric acid (>99.5%) and glycine (≥99.0%, NT) were purchased from Sigma-Aldrich (St. Louis, MO, USA). Sodium phosphate dibasic (ACS Grade), HCl (ACS Grade), and NaOH (ACS Grade) were all purchased from Fisher Scientific (Suwanee, GA, USA). Human apo-transferrin ("Apo-Tf") (≥95%) and human holo-Tf (≥95%) were purchased from Calbiochem (San Diego, CA, USA). Tricine (electrophoresis grade) was purchased from MP Biomedicals (Solon, OH, USA) and tris(hydroxymethyl)aminomethane (proteomics grade, Amresco Life Science, Solon, OH, USA) was purchased from VWR (Atlanta, GA, USA). Ultrapure water, purified using a Milli-Q® Reagent Water System from EMD Millipore Corporation (Billerica, MA, USA), was used for all aqueous samples and solutions.

3.2. Carbon Dots

The CDs used in this work were prepared in-house following the method from Dong et al. [6] with some modifications. Briefly, 2 g of dry citric acid in a 20-mL disposable scintillation vial was heated in an Isotemp oven (model 506G; Fisher Scientific, Waltham, MA, USA) at 180 °C for four hours. Alternatively, 2 g of dry citric acid was placed in a 50 mL Teflon autoclave liner, which was placed into a standard stainless steel 304 autoclave reactor (purchased from Labware on Amazon.com, part number: 2T50, Wilmington, DE, USA) and heated in the oven at 180 °C for 24 hours. The resulting dark orange liquid was cooled slightly and a 20 mL aqueous solution of 50 mM NaOH was added to the scintillation vial (or autoclave reactor) and was sonicated using a 2510 Branson sonicator (Branson Ultrasonics Corporation, Danbury, CT, USA) for 30 min to suspend the CDs. Three, 0.5-mL aliquots of the resulting neat CD solutions were lyophilized with a FreeZone 2.5 Liter −84 °C Benchtop Freeze Dryer (Labconco, Kansas City, MO, USA), and the mass of the resulting dried product was found. The average mass of three dried CD aliquots was found in order to provide a representative mass-per-volume (mg/mL) concentration of CDs for the batch. In some cases, a post-synthesis cleanup was performed by dialyzing (Float-a-Lyzer G2, MWCO 500-1000 D, from Spectrum Labs, (purchased from Fisher Scientific, Suwanee, GA, USA)) about 5 mL of the neat CD solution against water for eight hours, changing the water every two hours.

3.3. Separation Buffer and Sample Preparation

Stock solutions (1.00 M) of each buffer component (tris and tricine) were prepared separately by dissolving the appropriate mass of reagent in water, quantitatively transferring to a volumetric flask and filling to the line with ultrapure water. The resulting stock solutions were filtered (0.2 μm nylon syringe filter, VWR) and stored in a polypropylene or HDPE vessel at 2–8 °C until needed. The stock solutions were brought to room temperature before use. The tris-tricine buffer used for fluorescence emission and CE studies, was prepared to a final concentration of 50.0 mM tris and 200.0 mM tricine (unadjusted pH 7.4).

Additionally, other separation buffers were prepared from phosphate, tris and glycine. The phosphate separation buffer was prepared to a final concentration of 10.0 mM dibasic sodium phosphate adjusted to pH 7.4 with 1.0 M phosphoric acid. Two different tris-glycine buffers were prepared, one with a final concentration of 50.0 mM tris-200.0 mM glycine, and the other with 50.0 mM tris-500.0 mM glycine, and both adjusted to pH 7.4 by the dropwise addition of 1.0 M HCl. Lastly, a tris-HCl separation buffer was prepared to a final concentration of 50.0 mM tris adjusted to pH 7.4 with 1.0 M HCl.

Separate apo- and holo-Tf stock solutions (250 μM each) were prepared by dissolving the 0.01 g of apo-Tf or holo-Tf in the 500 μL of filtered ultrapure water (0.2 μm, nylon syringe filter) in a 1.6 mL microcentrifuge vial. Unused Tf stock solutions (of apo- and holo-Tf, separately) were portioned into 5 μL aliquots and stored at −20 °C until needed.

Samples were prepared for analysis by adding the appropriate volumes of apo-Tf, holo-Tf, or both stock solution(s) to a 1.6 mL microcentrifuge vial (Fisher Scientific, Suwanee, GA, USA) for fluorimetry studies and a 0.6 mL microcentrifuge vial (Fisher Scientific) for CE studies, followed by dilution with the appropriate volume of buffer (50 mM tris-200 mM tricine pH 7.4) for fluorimetry measurements, and with the appropriate volumes of sample buffer (100 mM tris-400 mM tricine pH 7.4, with 70 μg/mL CDs when present) and ultrapure water (producing a sample with a total volume of 50 μL) such that the final buffer concentration was (50 mM tris-200 mM tricine pH 7.4) for CE samples. The final buffer concentrations for fluorimetry samples are shown in Table S-1. For fluorimetry studies, a working solution of CDs (350 μg/mL) was prepared each time the studies were performed, from the neat solution of CDs after sonication of the neat solution for 5 min followed by approximately 25-fold dilution of a 39.78 μL portion of the neat solution with 960.22 μL buffer. The working solution was sonicated for 1–2 min prior to transferring the appropriate volume to the microcentrifuge vial containing the diluted Tf protein immediately prior to analysis (producing a sample with total volume of 500 μL). For CE samples, neat CDs solution (a 25.6 µL aliquot) was diluted approximately 390 fold with separation buffer in a 10.00 mL volumetric flask (resulting in a separation buffer with a final concentration of 35 µg/mL CDs), and a 2.56 µL aliquot of neat CDs solution was diluted 195 fold with 497.44 µL sample buffer in a 1.6 mL microcentrifuge tube (resulting in sample buffer with a final concentration of 70 µg/mL CDs). The sample vial was then vortexed to mix all constituents, and the solution therein (containing various combinations of apo-Tf, holo-Tf, buffer, and CDs, as needed) was transferred to a clean, dry, injection vial (Agilent, 250 µL or Beckman, 200 µL) for CE studies or to a semi-micro quartz cuvette (Fisher Scientific) for fluorimetry studies. The cuvette was cleaned by triple-rinsing with water and with 95% ethanol (Fisher Scientific).

3.4. Instrumentation

Spectrofluorimetry studies were conducted using a Cary Eclipse fluorescence spectrophotometer (Agilent Technologies, Foster City, CA, USA). An excitation wavelength of 360 nm was used, followed by an emission scan from 365–700 nm. Excitation and emission slit widths were 5 nm; the scan rate was 300 nm/min; and the PMT voltage was 600 V. CE studies were conducted using a P/ACE MDQ CE System with 32Karat software (Beckman Coulter, Redwood City, CA, USA) or an Agilent G1600A CE System equipped with Chemstation software. Detection was performed by UV absorbance at 200 nm, or by laser-induced fluorescence (LIF) using a 375 nm diode laser with 5 mW output power (Oz Optics Ltd., Carp, ON, Canada) and 400 nm long pass filter (Omega Optical, Brattleboro, VT, USA), or a Picometrics LIF Detector (406 nm laser with 12.5 mW output power and 410 nm emission filter) for the Beckman-Coulter and Agilent CE systems, respectively. All CE experiments employed uncoated fused-silica capillaries (Polymicro Technologies, Phoenix, AZ, USA) with different lengths and inside diameters (as specified in the Results and Discussion section).

4. Conclusions

The use of CDs as separation adjuvants in CE method development is presented as an opportunity to expand upon the usual repertoire of pseudo-stationary phases and buffer additives for enhanced separations. CDs employed in this study were synthesized in-house by a simple method of oven pyrolysis of citric acid. It is of significance that CDs were found to interact differentially with the various forms of Tf protein (metallated, demetallated, and partially metallated), as evidenced by varying extents of fluorescence quenching (which occurred for holo-Tf but not apo-Tf), and by a much more pronounced change in electrophoretic mobility for apo-Tf relative to holo-Tf with CDs present in the sample and separation buffers. This differential association of CDs with metallated and demetallated proteins facilitated greater resolution of apo- and holo-Tf by CE, along with the added ability to discern an additional sample component in the resulting electropherograms, which is presumed to be a partially metallated form of the protein, arising from spontaneous metal ion exchange between holo-Tf and apo-Tf components of the sample. Specifically, by employing a 25 µm i.d. capillary at 15 °C with a 50 mM tris-200 mM tricine separation buffer (pH 7.4) containing 35 µg/mL CDs, we were able to resolve three peaks for a sample comprising 25 µM each of apo-Tf and holo-Tf with 35 µg/mL CDs in water. Most importantly, resolution of these sample components was not possible in the absence of CDs. These results indicate that CDs are useful as CE buffer additives and can lead to improved resolution for challenging samples such as metallated protein mixtures. The application of CDs to other separation challenges in CE is a promising avenue for future studies.

Improvements to the method presented herein include efforts to resolve the two partially metallated forms of Tf co-migrating as the middle peak in the separations of mixtures of apo- and holo-Tf with CDs synthesized from citric acid, through modifications of the CDs, such as the addition of nitrogen groups or passivation with polymer, further optimization of the CE separation of mixtures of apo- and holo-Tf involving CDs from *N*-acetylneuraminic acid or glucose, optimization of the separation voltage, and determining if the partially metallated form of Tf is eliminated or reduced by the addition of 20% methanol to the separation buffer, which was found help reduce the loss of Fe^{3+} ions form holo

Tf. Additionally, the benefits of CDs in a polymer enhanced capillary transient isotachophoresis (PectI) method for mixtures of apo- and holo-Tf will be investigated.

Supplementary Materials: The following are available online, Table S-1: Final Buffer Concentrations for Fluorimetry Samples; Table S-2: Resolution Values for Figure 4; Figure S-1: Fluorescence spectra of apo- and holo-Tf without CDs and with Autoclave CDs; Figures S-2–S-9, representative electropherograms for method development and optimization studies.

Author Contributions: H.C.M. conducted preliminary studies for carbon dot synthesis and conceived of the initial idea to employ carbon dots for analyte sensing. S.S. extended the initial idea to include transferrin as the target. L.R.S. designed and conducted synthesis, spectroscopic, and capillary electrophoresis experiments with guidance from H.C.M., S.S., and C.L.C. Data analysis was completed by L.R.S. under the supervision of C.L.C. and S.S. Resources were provided by C.L.C. and S.S. The manuscript was written by L.R.S. and C.L.C. with review and editorial contributions from S.S. and H.C.M. Data curation and funding acquisition were performed by C.L.C. and L.R.S.

Funding: Financial support for this work was provided in part by Wake Forest University and the National Science Foundation (EAPSI Fellowship (141489) and GOALI Grant (1611072)).

Acknowledgments: The authors would like thank Tom Wittmann, Gina Li, and Emily Walton for their contributions to some of the experimental precursors to this work.

Conflicts of Interest: The authors declare no conflict of interest. The funders had no role in the design of the study; in the collection, analyses, or interpretation of data; in the writing of the manuscript, or in the decision to publish the results.

References

1. Shan, D.; Hsieh, J.-T.; Bai, X.; Yang, J. Citrate-based fluorescent biomaterials. *Adv. Healthc. Mater.* **2018**, *7*. [CrossRef]

2. Zhao, P.; Zhu, L. Dispersibility of carbon dots in aqueous and/or organic solvents. *Chem. Commun.* **2018**, *54*, 5401–5406. [CrossRef] [PubMed]

3. Sun, Y.; Bi, Q.; Zhang, X.; Wang, L.; Zhang, X.; Dong, S.; Zhao, L. Graphene quantum dots as additives in capillary electrophoresis for separation cinnamic acid and its derivatives. *Anal. Biochem.* **2016**, *500*, 38–44. [CrossRef]

4. Miao, P.; Han, K.; Tang, Y.; Wang, B.; Lin, T.; Cheng, W. Recent advances in carbon nanodots: Synthesis, properties and biomedical applications. *Nanoscale* **2015**, *7*, 1586–1595. [CrossRef]

5. Ju, J.; Chen, W. Synthesis of highly fluorescent nitrogen-doped graphene quantum dots for sensitive, label-free detection of Fe (III) in aqueous media. *Biosens. Bioelectron.* **2014**, *58*, 219–225. [CrossRef]

6. Dong, Y.; Shao, J.; Chen, C.; Li, H.; Wang, R.; Chi, Y.; Lin, X.; Chen, G. Blue luminescent graphene quantum dots and graphene oxide prepared by tuning the carbonization degree of citric acid. *Carbon* **2012**, *50*, 4738–4743. [CrossRef]

7. Qu, S.; Wang, X.; Lu, Q.; Liu, X.; Wang, L. A biocompatible fluorescent ink based on water-soluble luminescent carbon nanodots. *Angew. Chem. Int. Ed.* **2012**, *51*, 12215–12218. [CrossRef] [PubMed]

8. Yang, W.; Zhang, H.; Lai, J.; Peng, X.; Hu, Y.; Gu, W.; Ye, L. Carbon dots with red-shifted photoluminescence by fluorine doping for optical bio-imaging. *Carbon* **2018**, *128*, 78–85. [CrossRef]

9. Bhaisare, M.L.; Talib, A.; Khan, M.S.; Pandey, S.; Wu, H.-F. Synthesis of fluorescent carbon dots via microwave carbonization of citric acid in presence of tetraoctylammonium ion, and their application to cellular bioimaging. *Microchim. Acta* **2015**, *182*, 2173–2181. [CrossRef]

10. Goh, E.J.; Kim, K.S.; Kim, Y.R.; Jung, H.S.; Beack, S.; Kong, W.H.; Scarcelli, G.; Yun, S.H.; Hahn, S.K. Bioimaging of hyaluronic acid derivatives using nanosized carbon dots. *Biomacromolecules* **2012**, *13*, 2554–2561. [CrossRef] [PubMed]

11. Zholobak, N.M.; Popov, A.L.; Shcherbakov, A.B.; Popova, N.R.; Guzyk, M.M.; Antonovich, V.P.; Yegorova, A.V.; Scrypynets, Y.V.; Leonenko, I.I.; Baranchikov, A.Y.; et al. Facile fabrication of luminescent organic dots by thermolysis of citric acid in urea melt, and their use for cell staining and polyelectrolyte microcapsule labelling. *Beilstein J. Nanotechnol.* **2016**, *7*, 1905–1917. [CrossRef]

12. Zheng, M.; Liu, S.; Li, J.; Qu, D.; Zhao, H.; Guan, X.; Hu, X.; Xie, Z.; Jing, X.; Sun, Z. Integrating oxaliplatin with highly luminescent carbon dots: An unprecedented theranostic agent for personalized medicine. *Adv. Mater.* **2014**, *26*, 3554–3560. [CrossRef] [PubMed]

13. Iqbal, A.; Iqbal, K.; Xu, L.; Li, B.; Gong, D.; Liu, X.; Guo, Y.; Liu, W.; Qin, W.; Guo, H. Heterogeneous synthesis of nitrogen-doped carbon dots prepared via anhydrous citric acid and melamine for selective and sensitive turn on-off-on detection of Hg (II), glutathione and its cellular imaging. *Sens. Actuators B Chem.* **2018**, *255*, 1130–1138.

14. Zhang, Q.; Zhang, C.; Li, Z.; Ge, J.; Li, C.; Dong, C.; Shuang, S. Nitrogen-doped carbon dots as fluorescent probe for detection of curcumin based on the inner filter effect. *RSC Adv.* **2015**, *5*, 95054–95060. [CrossRef]

15. Wang, Y.; Gao, D.; Chen, Y.; Hu, G.; Liu, H.; Jiang, Y. Development of N, S-doped carbon dots as a novel matrix for the analysis of small molecules by negative ion MALDI-TOF MS. *RSC Adv.* **2016**, *6*, 79043–79049. [CrossRef]

16. Duan, L.-P.; Ding, G.-S.; Tang, A.-N. Preparation of chitosan-modified silica nanoparticles and their applications in the separation of auxins by capillary electrophoresis. *J. Sep. Sci.* **2015**, *38*, 3976–3982. [CrossRef]

17. Gong, Z.-S.; Duan, L.-P.; Tang, A.-N. Amino-functionalized silica nanoparticles for improved enantiomeric separation in capillary electrophoresis using carboxymethyl-beta-cyclodextrin (CM-beta-CD) as a chiral selector. *Microchim. Acta* **2015**, *182*, 1297–1304. [CrossRef]

18. Manuel Jimenez-Soto, J.; Moliner-Martinez, Y.; Cardenas, S.; Valcarcel, M. Evaluation of the performance of singlewalled carbon nanohorns in capillary electrophoresis. *Electrophoresis* **2010**, *31*, 1681–1688. [CrossRef]

19. Benitez-Martinez, S.; Simonet, B.M.; Valcarcel, M. Graphene nanoparticles as pseudostationary phase for the electrokinetic separation of nonsteroidal anti-inflammatory drugs. *Electrophoresis* **2013**, *34*, 2561–2567. [CrossRef]

20. Cao, J.; Qu, H.; Cheng, Y. Separation of flavonoids and phenolic acids in complex natural products by microemulsion electrokinetic chromatography using surfactant-coated and carboxylic single-wall carbon nanotubes as additives. *Electrophoresis* **2010**, *31*, 1689–1696. [CrossRef]

21. Neiman, B.; Grushka, E.; Lev, O. Use of gold nanoparticles to enhance capillary electrophoresis. *Anal. Chem.* **2001**, *73*, 5220–5227. [CrossRef] [PubMed]

22. Viberg, P.; Jornten-Karlsson, M.; Petersson, P.; Spegel, P.; Nilsson, S. Nanoparticles as pseudostationary phase in capillary electrochrornatography/ESI-MS. *Anal. Chem.* **2002**, *74*, 4595–4601. [CrossRef]

23. Zhao, T.; Zhou, G.; Wu, Y.; Liu, X.; Wang, F. Gold nanomaterials based pseudostationary phases in capillary electrophoresis: A brand-new attempt at chondroitin sulfate isomers separation. *Electrophoresis* **2015**, *36*, 588–595. [CrossRef]

24. Harris, D.C. Chromatographic methods and capillary electrophoresis. In *Quantitative Chemical Analysis*; W.H. Freeman and Co.: New York, NY, USA, 2003; pp. 654–672. ISBN 0-7167-4464-3.

25. Kinsel, G.R. Miscellaneous separation methods. In *Fundamentals of Analytical Chemistry 8th EDITION*; Skoog, D.A., West, D.M., Holler, F.J., Crouch, S.R., Eds.; Cengage Learning: Boston, MA, USA, 2003.

26. Nowak, P.; Spiewak, K.; Brindell, M.; Wozniakiewicz, M.; Stochel, G.; Koscielniak, P. Separation of iron-free and iron-saturated forms of transferrin and lactoferrin via capillary electrophoresis performed in fused-silica and neutral capillaries. *J. Chromatogr. A* **2013**, *1321*, 127–132. [CrossRef]

27. Yue, C.-Y.; Ding, G.-S.; Liu, F.-J.; Tang, A.-N. Water-compatible surface molecularly imprinted silica nanoparticles as pseudostationary phase in electrokinetic chromatography for the enantioseparation of tryptophan. *J. Chromatogr. A* **2013**, *1311*, 176–182. [CrossRef]

28. Turiel, E.; Martin-Esteban, A. Molecular imprinting technology in capillary electrochromatography. *J. Sep. Sci.* **2005**, *28*, 719–728. [CrossRef] [PubMed]

29. Ren, L.; Kim, H.K.; Zhong, W. Capillary electrophoresis-assisted identification of peroxyl radical generated by single-walled carbon nanotubes in a cell-free system. *Anal. Chem.* **2009**, *81*, 5510–5516. [CrossRef]

30. Lippard, S.J.; Berg, J.M. *Principles of Bioinorganic Chemistry*; University Science Books: Sausalito, CA, USA, 1994; ISBN 0-935702-72-5.

31. Li, H.Y.; Qian, Z.M. Transferrin/transferrin receptor-mediated drug delivery. *Med. Res. Rev.* **2002**, *22*, 225–250. [CrossRef]

32. Ali, S.A.; Joao, H.C.; Hammerschmid, F.; Eder, J.; Steinkasserer, A. Transferrin Trojan horses as a rational approach for the biological delivery of therapeutic peptide domains. *J. Biol. Chem.* **1999**, *274*, 24066–24073. [CrossRef] [PubMed]

33. Singh, M. Transferrin as a targeting ligand for liposomes and anticancer drugs. *Curr. Pharm. Des.* **1999**, *5*, 443–451.

34. Qian, Z.M.; Li, H.Y.; Sun, H.Z.; Ho, K. Targeted drug delivery via the transferrin receptor-mediated endocytosis pathway. *Pharmacol. Rev.* **2002**, *54*, 561–587. [CrossRef]

35. Du, W.; Fan, Y.; Zheng, N.; He, B.; Yuan, L.; Zhang, H.; Wang, X.; Wang, J.; Zhang, X.; Zhang, Q. Transferrin receptor specific nanocarriers conjugated with functional 7peptide for oral drug delivery. *Biomaterials* **2013**, *34*, 794–806. [CrossRef]

36. Camp, E.R.; Wang, C.; Little, E.C.; Watson, P.M.; Pirollo, K.F.; Rait, A.; Cole, D.J.; Chang, E.H.; Watson, D.K. Transferrin receptor targeting nanomedicine delivering wild-type p53 gene sensitizes pancreatic cancer to gemcitabine therapy. *Cancer Gene Ther.* **2013**, *20*, 222–228. [CrossRef]

37. Martinez, A.; Suarez, J.; Shand, T.; Magliozzo, R.S.; Sanchez-Delgado, R.A. Interactions of arene-Ru(II)-chloroquine complexes of known antimalarial and antitumor activity with human serum albumin (HSA) and transferrin. *J. Inorg. Biochem.* **2011**, *105*, 39–45. [CrossRef]

38. Tortorella, S.; Karagiannis, T.C. Transferrin receptor-mediated endocytosis: A useful target for cancer therapy. *J. Membr. Biol.* **2014**, *247*, 291–307. [CrossRef] [PubMed]

39. Bhattacharya, A.; Chatterjee, S.; Khorwal, V.; Mukherjee, T.K. Luminescence turn-on/off sensing of biological iron by carbon dots in transferrin. *Phys. Chem. Chem. Phys.* **2016**, *18*, 5148–5158. [CrossRef]

40. Nowak, P.; Spiewak, K.; Nowak, J.; Brindell, M.; Wozniakiewicz, M.; Stochel, G.; Koscielniak, P. Selective separation of ferric and non-ferric forms of human transferrin by capillary micellar electrokinetic chromatography. *J. Chromatogr. A* **2014**, *1341*, 73–78. [CrossRef] [PubMed]

41. Zhu, S.; Meng, Q.; Wang, L.; Zhang, J.; Song, Y.; Jin, H.; Zhang, K.; Sun, H.; Wang, H.; Yang, B. Highly photoluminescent carbon dots for multicolor patterning, sensors, and bioimaging. *Angew. Chem. Int. Ed.* **2013**, *52*, 3953–3957. [CrossRef] [PubMed]

![molecules logo]

molecules

MDPI

Article

Identification of a Recombinant Human Interleukin-12 (rhIL-12) Fragment in Non-Reduced SDS-PAGE

Lei Yu [†], Yonghong Li [†], Lei Tao [†], Chuncui Jia, Wenrong Yao, Chunming Rao * and Junzhi Wang *

National Institutes for Food and Drug Control, Beijing 100050, China; yulei@nifdc.org.cn (L.Y.);
liyh@nifdc.org.cn (Y.L.); taolei01@nifdc.org.cn (L.T.); chuncui319@163.com (C.J.); yz1322@126.com (W.Y.)
* Correspondence: raocm@nifdc.org.cn (C.R.); wangjz@nifdc.org.cn (J.W.);
  Tel.: +86-106-709-5586 (J.W.); Fax: +86-106-701-8094 (J.W.)
† These authors contributed equally to this work.

Academic Editors: Angela R. Piergiovanni and José Manuel Herrero-Martínez
Received: 18 February 2019; Accepted: 23 March 2019; Published: 28 March 2019

Abstract: During the past two decades, recombinant human interleukin-12 (rhIL-12) has emerged as one of the most potent cytokines in mediating antitumor activity in a variety of preclinical models and clinical studies. Purity is a critical quality attribute (CQA) in the quality control system of rhIL-12. In our study, rhIL-12 bulks from manufacturer B showed a different pattern in non-reduced SDS-PAGE compared with size-exclusion chromatography (SEC)-HPLC. A small fragment was only detected in non-reduced SDS-PAGE but not in SEC-HPLC. The results of UPLC/MS and N-terminal sequencing confirmed that the small fragment was a 261–306 amino acid sequence of a p40 subunit of IL-12. The cleavage occurs between Lys260 and Arg261, a basic rich region. With the presence of 0.2% SDS, the small fragment appeared in both native PAGE and in SEC-HPLC, suggesting that it is bound to the remaining part of the IL-12 non-covalently, and is dissociated in a denatured environment. The results of a bioassay showed that the fractured rhIL-12 proteins had deficient biological activity. These findings provide an important reference for the quality control of the production process and the final products of rhIL-12.

Keywords: rhIL-12; purity; SDS-PAGE; SEC-HPLC; fragment; non-covalent binding

---

1. Introduction

Interleukin-12 (IL-12) is a key inflammatory cytokine that critically influences Th1/Tc1-T cell responses at the time of an initial antigen encounter [1,2]. A growing number of studies have shed light on its potential in cancer immunotherapy [3,4]. During the past two decades, IL-12 has emerged as one of the most potent cytokines in mediating antitumor activity in a variety of preclinical models and clinical studies [5–7]. IL-12 has multiple biological functions, though most importantly, it bridges early nonspecific innate resistance and the subsequent antigen-specific adaptive immunity. A remarkable function of IL-12 is its ability to induce interferon γ (IFNγ) release from natural killer (NK) cells as well as CD4+ and CD8+ T cells [8,9]. Increasingly, pharmaceutical companies have invested in the research and development of recombinant human IL-12 (rhIL-12) [5,10,11]. One such company in China has obtained a national first class clinical approval for rhIL-12 injection. Aside from this, some IL-12 fusion proteins were developed to minimize adverse effects, such as huBC1-IL-12, F8-IL-12 and IL-12-SS1 (Fv), as well as IL-12 gene therapy products [8,11–13].

Native IL-12 is a heterodimer formed by two subunits, p40 (306 amino acids) and p35 (197 amino acids), that are bridged by an inter-subunit disulfide bond between Cys177 of p40 and Cys74 of p35, with three potential N-glycosylation sites [14,15]. Genes of the subunits p40 and p35 were located in different chromosomes in the human genome. The common preparation strategy for rhIL-12 is that *p40* and *p35* cDNA sequences were inserted into either two different vectors or one vector with two different promoters, and then transfected into a mammalian expression system [5,10]. As a recombinant protein drug, purity is a critical quality attribute (CQA) in the quality control system of rhIL-12. A purity test is essential for lot release and stability testing. The conventional analytical methods include chromatography and electrophoresis methods, such as size-exclusion chromatography (SEC), ion-exchange chromatography (IEC), denaturing protein gel electrophoresis (SDS-PAGE), capillary electrophoresis (CE)-SDS and capillary isoelectric focusing (cIEF) [16]. Aside from this, a liquid chromatography-mass spectrometry (LC-MS)-based multi-attribute method (MAM) has recently become a research hotspot [17,18]. As stated in the ICH Q6B, 'the determination of absolute, as well as relative purity, presents considerable analytical challenges, and the results are highly method-development.' Therefore, the purity must be assessed by a combination of analytical procedures. For most recombinant protein drugs, a combination of SDS-PAGE and SEC-HPLC are recommended. In our study, the purity of rhIL-12 bulks from manufacturer B was determined by non-reduced SDS-PAGE and SEC-HPLC, but the results were inconsistent. A small fragment was detected in non-reduced SDS-PAGE but not in SEC-HPLC. We used UPLC/MS and N-terminal sequencing to identify the fragment and then attempted to find out the cause of the cleavage and its effect on biological activity.

2. Results and Discussion

2.1. Purity Determination of rhIL-12 Samples by Non-Reduced SDS-PAGE and SEC-HPLC

Three batches of rhIL-12 bulks (S01, S02 and S03) from manufacturer B were tested by non-reduced SDS-PAGE and SEC-HPLC. The electrophoretogram and chromatogram are shown in Figure 1A,B, and the relative percentage contents are listed in Table 1. High molecular proteins, generally known as protein multimers, were detected in both assays, but the relative percentage contents in SEC-HPLC were significantly higher than those in SDS-PAGE, which may be caused by the different running system—native for SEC-HPLC and denatured for SDS-PAGE. The denaturation led most of the non-covalent multimers to be depolymerized, so the multimers in SDS-PAGE were generally significantly lower than in SEC-HPLC. However, fragments were only detected in non-reduced SDS-PAGE, and the relative percentage contents exceeded 7%. It was necessary to figure out the component of the small fragment present in non-reduced SDS-PAGE but absent in SEC-HPLC, as well as its origin and whether it was produced during the production process or during the testing process. No obvious small fragment was found in non-reduced SDS-PAGE for an rhIL-12 in-house reference (Figure 1A), suggesting that the fragment was not produced during SDS-PAGE. All test samples were bulks without any ingredients (such as a stabilizer) and were stored at −70 °C since prepared, which was conducted for over two years for the in-house reference and for a few months for the S01, S02 and S03 batches. It is generally recognized that proteins should be stable at −70 °C for an extended period of time (ultimately for a period of years). The in-house reference in this case had been stored for an even longer period, suggesting that the small fragment was not produced during storage. Thus it could be a product-related impurity or a process-related impurity originally existing in the rhIL-12 bulks. As an unknown protein impurity, it may bring about safety risks (such as toxicity or immunogenicity). We tried to identify the small fragment in SDS-PAGE in the following study.

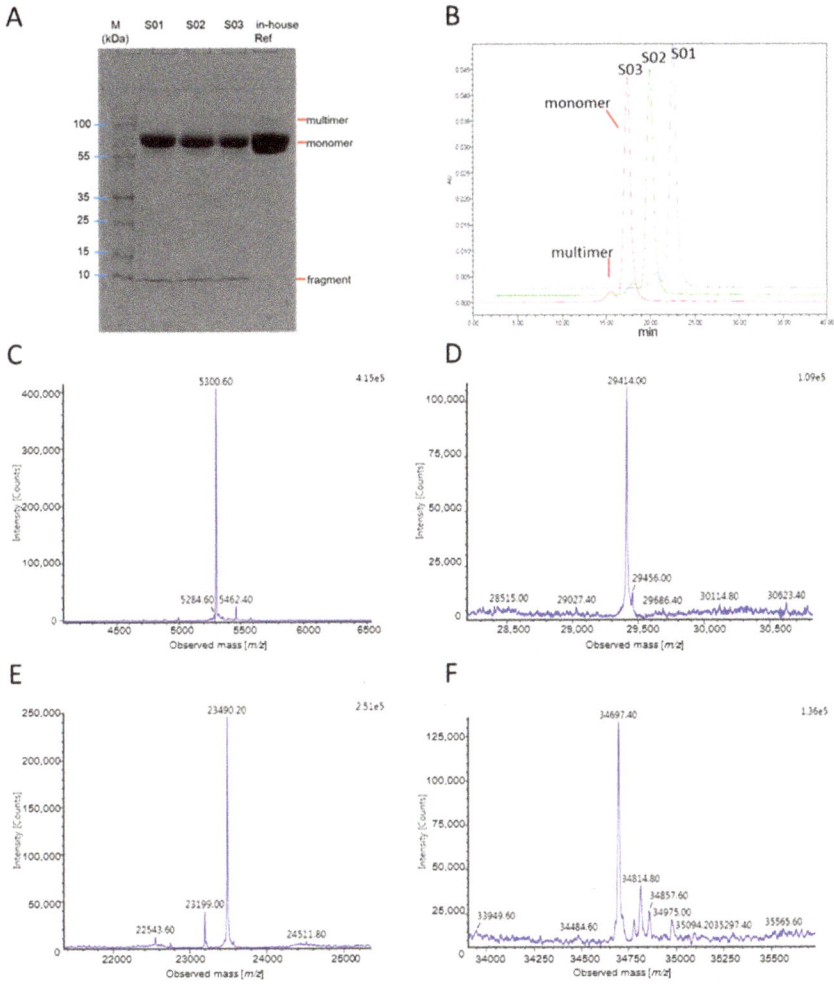

Figure 1. Purity and molecular weight determination of rhIL-12. (**A,B**) Purity determination of rhIL-12 bulks (S01, S02 and S03) by non-reduced SDS-PAGE and SEC-HPLC. (**C,D**) Molecular weight of p40 fragments by MS for rhIL-12 sample S01. (**E**) Molecular weight of p35 by MS for rhIL-12 sample S01. (**F**) Molecular weight of intact p40 by MS for rhIL-12 in-house reference.

Table 1. Purity of recombinant human interleukin-12 (rhIL-12) bulks by SDS-PAGE and size-exclusion chromatography (SEC)-HPLC.

Sample ID	Multimer (%) [a]		Monomer (%) [a]		Fragment (%) [a]	
	SDS-PAGE	**SEC-HPLC**	**SDS-PAGE**	**SEC-HPLC**	**SDS-PAGE**	**SEC-HPLC**
S01	0.69	4.19	92.00	95.81	7.31	_[b]
S02	0.71	2.96	91.32	97.04	7.97	_[b]
S03	0.98	4.60	91.29	95.40	7.73	_[b]

[a]. The relative percentage contents were calculated by the area normalization method. [b]. Not detected.

2.2. Identification of rhIL-12 Fragment by UPLC/MS and N-Terminal Sequencing

UPLC/MS was employed to detect if there was any cleavage in the rhIL-12 peptides. For most glycoproteins, N-linked sugar chains are complex and heterogeneous, and are always removed for the determination of molecular weight (MW) by MS. The rhIL-12 samples were denatured by dithiothreitol (DTT) and deglycosylated by PNGase F before analysis. MW results are listed in Table 2. For subunit p35, although O-glycosylation caused heterogeneity in its MW, the measured and theoretical MWs were basically matched (Figure 1E). As for subunit p40, it is composed of 306 amino acids and its theoretical MW is 34698.03 Da. The measured value of the rhIL-12 in-house reference was 34697.40 Da (Figure 1F), which is highly consistent with the theoretical value. However, no intact p40 subunit was found in the rhIL-12 sample from manufacturer B. Instead, two fragments of 5.3 kDa and 29.4 kDa were detected (Figure 1C,D). The 29.4 kDa and 5.3 kDa fragments were consistent with the 1–260 amino acid sequence and the 261–306 amino acid sequence of subunit p40, respectively, which suggests that a cleavage did occur in subunit p40, and that the cleavage site was between Lys260 and Arg261, a dibasic site, as listed in Table 2. Since the inter-subunit disulfide bond is between the Cys177 of subunit p40 and the Cys74 of subunit p35 [14], the 29.4 kDa fragment should still be linked to subunit p35 covalently.

Table 2. Molecular weight (MW) of rhIL-12 (S01) by MS.

Subunit	Amino Acid Sequence	Theoretical MW (Da)	Measured MW (Da)	Error (Da)	Relative Error (ppm)
p40	1–260	29415.12 [a]	29414.00	1.12	38
	261–306	5300.93	5300.60	0.33	62
		22544.21 [b]	22543.60	0.61	27
p35	1–197	23200.80 [c]	23199.00	1.80	78
		23492.06 [d]	23490.20	1.86	79

[a]. One N-glycosylation (Glu→Gln) caused an increase of 0.98 Da; [b]. Two N-glycosylations (Glu→Gln) caused an increase of 1.96 Da; [c]. O-glycosylation (GlcNAc-Man-SA) caused an increase of 656.59 Da based on b; [d]. O-glycosylation (GlcNAc-Man-2SA) caused an increase of 947.85 Da based on b.

To verify whether the small peptide in the non-reduced SDS-PAGE was the 5.3 kDa fragment of subunit p40, we further identified it by N-terminal sequencing. The measured sequence of 16 N-terminal amino acids was REKKDRVFTDKTSATV, which was completely consistent with the theoretical N-terminal sequence of the 5.3 kDa fragment, confirming that the small peptide in non-reduced SDS-PAGE was the 5.3 kDa fragment of subunit p40. The Cycles 2 to 6 are shown in Figure S1. The reason why it was only present in non-reduced SDS-PAGE but absent in SEC-HPLC may be due to the different running systems of SEC-HPLC (native) compared with SDS-PAGE (denatured). The fragment may bind to the remaining part of IL-12 non-covalently in a native environment but dissociate in a denatured environment. To test this further, we next evaluated the effect of the denaturant.

2.3. Effect of Denaturant on rhIL-12 Pattern in Native PAGE and SEC-HPLC

The rhIL-12 sample S01 was treated with 0%, 0.2% or 0.02% SDS for 10 min at room temperature before being tested by native PAGE and SEC-HPLC. In non-denatured PAGE, no fragment was found without the presence of SDS or with the presence of 0.02% SDS, but in the case where 0.2% SDS was employed, the small fragment reappeared (Figure 2A). In SEC-HPLC, with the presence of SDS (both 0.02% and 0.2%), the small fragment appeared (Figure 2B). SDS is amphipathic in nature, which allows it to unfold both polar and nonpolar sections of a protein structure. In SDS concentrations above 0.1 mM (0.003%), the unfolding of proteins begins, and above 1 mM (0.03%), most proteins are denatured [19]. These results suggest that the 5.3 kDa fragment bound to the remaining part of IL-12 non-covalently in a non-denatured environment, and dissociated with the presence of SDS. This explains why the 5.3 kDa fragment was only visible in SDS-PAGE but not in SEC-HPLC.

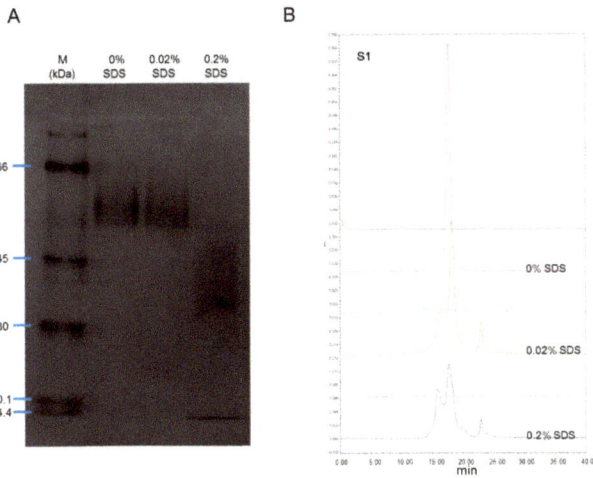

Figure 2. Effect of SDS on rhIL-12 pattern in native PAGE and SEC-HPLC. The rhIL-12 bulk S01 was treated by 0%, 0.02% and 0.2% SDS before analysis. (**A**) Native PAGE. (**B**) SEC-HPLC.

As for the inconsistency between native PAGE and SEC-HPLC with the presence of 0.02% SDS, this may be caused by a different reaction time. For PAGE, all samples were loaded onto the gel at the same time, but for SEC-HPLC, samples were analyzed in sequence (0%, 0.02% and 0.2% SDS), meaning that the reaction time of samples for SEC-HPLC were unequal and longer than for PAGE. The longer reaction time of SDS always brings about greater efficiency in terms of denaturation.

2.4. Cleavage Site in 3D Structure of rhIL-12

The p40 subunit consists of three domains, FN3, rhIL-12p40_C and IGc2 [20]. As shown in Figure 3, the cleavage occurs between Lys260 and Arg261, and a 5.3 kDa fragment is located in the FN3 domain of subunit p40, which is tightly folded. The cleavage site is a basic rich sequence (Lys-Ser-Lys-Arg-Glu-Lys-Lys) exposed on the surface of the molecule. Lys260-Arg261 is a dibasic site, liable to be targeted by endogenous protease [21–23], and residual proteases used for the removal of protein purification tag(s) (if any) may have a nonspecific effect on the site. Other than this, low pH conditions could also induce the instability of basic amino acids. The cleavage may occur during the production or the purification process. Intermediate products at different steps of the process should be analyzed to figure out at which step the cleavage occurred and to develop an appropriate strategy to avoid this occurrence.

In the spatial structure of IL-12, the 5.3 kDa sequence is located in the C-terminal of subunit p40, and binds tightly with the rest of the FN3 domain, which should be the reason for its absence in native PAGE and SEC-HPLC. However, in non-reduced SDS-PAGE, a denatured environment destroyed the non-covalent bond and the fragment was disassociated and finally appeared in the electrophoretogram. For proteins, non-reduced SDS-PAGE is usually the first choice as an assay of purity, not only because of its reliability and ease, but also because of its ability to separate the fragments binding to the principal component non-covalently.

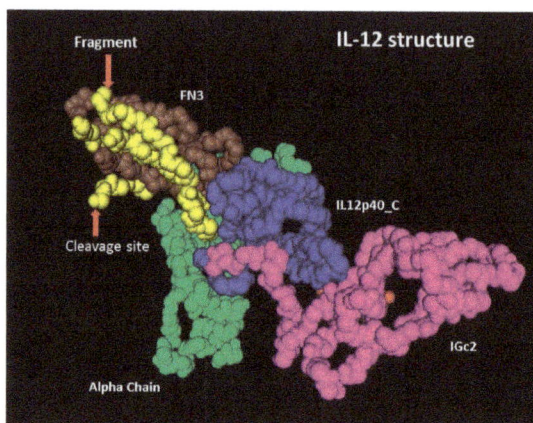

Figure 3. 3D structure of IL-12. The alpha chain (green) is subunit p35. Subunit p40 consists of three domains: FN3 (brown), rhIL-12p40_C (blue) and IGc2 (pink). The yellow region is the 5.3 kDa fragment, a part of the FN3 domain. This 3D structure was obtained from the Protein Data Bank (PDB) website (ID: 1F45).

2.5. Influence of Cleavage on Bioactivity

Although cleavage occurred, if the spatial structure remained intact, proteins could still function properly. Next, we confirmed whether this cleavage had any negative influence on its bioactivity by comparing its specific activity with the rhIL-12 in-house reference (intact IL-12). The biological activity of rhIL-12 was determined by NK92MI/interferon γ release assay, which served to test its induction of interferon γ release in NK92MI cells. Figure 4 shows the dose-response curves of the World Health Organization (WHO) biological standard for rhIL-12, the rhIL-12 in-house reference and samples (S01–S03). The results of the protein content and biological activity are listed in Table 3. Three batches of fractured rhIL-12 (S01, S02 and S03) showed half the specific activity of intact rhIL-12.

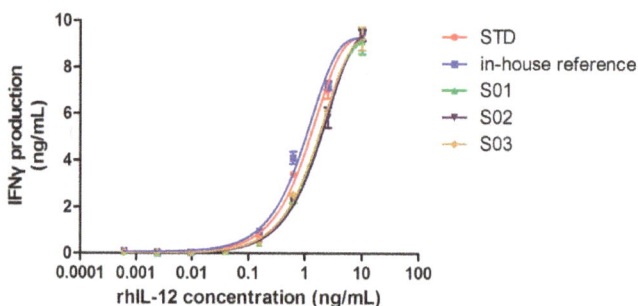

Figure 4. Dose-response curves of WHO biological standard for rhIL-12, the rhIL-12 in-house reference and rhIL-12 samples (S01–S03). Each plot represents the mean of two replicates.

Table 3. The results of protein content, biological activity and specific activity.

Samples	Protein Content (mg/mL, Mean of Three Replicates)	Biological Activity (units/mL, Mean of Three Replicates)	Specific Activity (units/mg)
In-house reference	1.78	1.57×10^7	8.81×10^6
S01	0.35	1.56×10^6	4.45×10^6
S02	0.34	1.46×10^6	4.31×10^6
S03	0.35	1.57×10^6	4.48×10^6

Although no intact IL-12 molecule was found in the rhIL-12 bulks from manufacturer B by MS, 50% of the total activities were reserved. The reason for this may be that half of the 5.3 kDa fragments were folded properly with the remaining part of the IL-12, forming an intact spatial structure, or that the incomplete spatial structure still retained partial activity, which should be further studied by spatial structure analysis. Nonetheless, the cleavage had a negative effect on the biological activity of rhIL-12.

3. Materials and Methods

3.1. Materials

The WHO biological standard for IL-12 was obtained from the National Institute for Biological Standards and Control (NIBSC code: 95/544). The rhIL-12 in-house reference (bulk from manufacturer A, Qingdao, China) and samples S01, S02 and S03 (different batches of bulks from manufacturer B, Guangzhou, China) were archived samples that had been preserved at −70 °C in our laboratory.

3.2. Electrophoresis Analysis

Purity was evaluated by non-reduced SDS-PAGE performed on a 4–20% SDS-tris-glycine gel (Thermo Fisher Scientific, Carlsbad, CA, USA). Samples were diluted in a non-reducing SDS sample buffer and heated at 95 ± 5 °C for 5 min with 10 μg of each sample loaded onto the gel. The samples were separated by electrophoresis and the gel was stained with 0.25% w/v Coomassie R-250 (Bio-Rad, Hercules, CA, USA), destained for clarity, and scanned. The relative percentage contents were calculated using the area normalization method. For native PAGE, SDS was excluded from the electrophoresis system.

3.3. Size-Exclusion Chromatography Analysis

LC separation was performed on a Waters2695 system with a TSK-GEL G3000 SWXL column (300 mm × 7.8 mm, Tosoh, Japan). The injection volume was 50 μL. The flow rate was 0.5 mL/min using an elution buffer of 40 mM phosphate buffer containing 300 mM sodium sulfate (pH 7.2) and the column temperature was maintained at 25 °C. The detection was performed on a Waters2489 UV detector (Waters, Milford, MA, USA) at 280 nm. Data were acquired and processed using Waters Empower (Waters Corporation, Milford, MA, USA). The relative percentage contents were calculated by the area normalization method.

3.4. UPLC/MS

The rhIL-12 samples were denatured by 10mM DTT (Sigma, St. Louis, MO, USA) and deglycosylated by PNGase F (New England Biolabs, Beijing, China), then analyzed by the Acquity UPLC system connected online to a Xevo G2-S mass spectrometer (Waters Corporation, Milford, MA, USA). The column was a Waters BEH300 C4 column (2.1 mm × 50 mm, 1.7 μm particle). The flow rate was 0.2 mL/min using a gradient from 5% to 50% Solvent B (Solvent B being 0.1% formic acid in acetonitrile, Solvent A being 0.1% formic acid in water) in 7 min at a column temperature of 35 °C. The scan range of the mass spectrometric was m/z 500–3000. Data were acquired and processed using UNIFI 1.6 (Waters Corporation, Milford, MA, USA).

3.5. N-Terminal Sequencing

The rhIL-12 sample (S01) was condensed to about 2 mg/mL by ultrafiltration using 3 kDa centrifugal filters (Merck Millipore Ltd., Tullagreen, Ireland). Then, 32 μL of the sample (64 μg) was mixed with 8 μL non-reducing SDS sample buffer (5×) and heated at 95 ± 5 °C for 5 min, and subsequently loaded onto a 4–20% Tris-glycine gel and separated by SDS-PAGE. Proteins were transferred electrophoretically onto a polyvinylidene fluoride (PVDF) membrane. The small fragment was excised and subjected to 16 cycles of N-terminal sequence analysis using a PPSQ-53A protein sequencer (Shimadzu, Kyoto, Japan).

3.6. Measurement of rhIL-12 Concentration and Bioactivity

The protein contents of the rhIL-12 in-house reference and samples were determined by a Pierce™ BCA protein assay kit (Thermo scientific, Rockford, IL, USA) according to the manufacturer's instructions. The bioactivity of rhIL-12 was determined by quantifying IFN-γ secretion by the IL-12-responsive NK-92MI cell line (American Type Culture Collection, Manassas, VA, USA) cultured in complete media consisting of Minimum Essential Medium α (MEMα) supplemented with 12% fetal bovine serum (FBS), 12% horse serum, 1% penicillin/streptomycin, 0.2 mM inositol, 0.02 mM folic acid and 0.1 mM 2-mercaptoethanol. In brief, cultured NK-92MI cells were seeded in a 96-well plate at 20,000 cells/well. The WHO biological standard for IL-12, the rhIL-12 in-house reference and the samples were added to the cells at final concentrations of 10 ng/mL~0.0006 ng/mL. IFN-γ concentrations in NK92-MI supernatants after 24 hours were quantified using an IFN-γ ELISA kit (BD Biosciences, Franklin Lakes, NJ, USA) according to the manufacturer's instructions.

4. Conclusions

Collectively, all our results proved that the small fragment in non-reduced SDS-PAGE was a 261–306 amino acid sequence of subunit p40, located in the FN3 domain, a tightly folded domain in the IL-12 spatial structure. The fragment could bind to the remaining part of IL-12 non-covalently and dissociate in a denatured environment. The cleavage site was found to be a basic rich sequence exposed on the surface of the molecule. The cleavage may occur during the production or the purification processes. Fractured rhIL-12 proteins had deficient biological activity. Additionally, the cleavage was not unique, found not only in samples from manufacturer B but also in samples from another manufacturer, manufacturer C (data not shown). Thus, it is necessary to find the cause of the cleavage and develop an appropriate strategy to avoid its occurrence. Our work provides an important reference for the quality control of the production process and final products of rhIL-12, as well as an improvement in the production technology. This study reveals the importance of purity determination through a combination of analytical procedures with different principles.

Supplementary Materials: The following are available online, Figure S1: Determination of N-terminal amino acid sequence by Edman degradation.

Author Contributions: Conceptualization, L.Y., C.R. and J.W.; methodology, L.Y., Y.L. and L.T.; software, L.Y.; validation, L.T. and C.J.; formal analysis, Y.L.; investigation, L.T.; resources, C.R.; data curation, L.Y.; writing—original draft preparation, L.Y.; writing—review and editing, C.R. and W.Y.; visualization, L.T.; supervision, C.R. and J.W.; project administration, J.W.; funding acquisition, L.Y. and J.W.

Funding: This work was financially supported by grants from the National Science and Technology Major Project (grant number 2018ZX09101001) and the Middle-Aged and Young Development Research Foundation of NIFDC (No. 2017B3). The funders had no role in study design, data collection and analysis, decision to publish, or preparation of the manuscript.

Conflicts of Interest: The authors declare no conflict of interest.

References

1. Hamza, T.; Barnett, J.B.; Li, B. Interleukin 12 a key immunoregulatory cytokine in infection applications. *Int. J. Mol. Sci.* **2010**, *11*, 789–806. [CrossRef]
2. Sun, L.; He, C.; Nair, L.; Yeung, J.; Egwuagu, C.E. Interleukin 12 (IL-12) family cytokines: Role in immune pathogenesis and treatment of CNS autoimmune disease. *Cytokine* **2015**, *75*, 249–255. [CrossRef]
3. Braun, M.; Ress, M.L.; Yoo, Y.E.; Scholz, C.J.; Eyrich, M.; Schlegel, P.G.; Wolfl, M. IL12-mediated sensitizing of T-cell receptor-dependent and -independent tumor cell killing. *Oncoimmunology* **2016**, *5*, e1188245. [CrossRef]
4. Del Vecchio, M.; Bajetta, E.; Canova, S.; Lotze, M.T.; Wesa, A.; Parmiani, G.; Anichini, A. Interleukin-12: Biological properties and clinical application. *Clin. Cancer Res.* **2007**, *13*, 4677–4685. [CrossRef] [PubMed]

5. Jiao, H.Y.; Zhan, M.Y.; Guo, M.Z.; Yi, Y.; Cong, Y.; Tian, R.G.; Zhang, W.Y.; Bi, S.L. Expression of human IL-12 in mammalian cell and study on its biological activities. *Chin. J. Exp. Clin. Virol.* **2007**, *21*, 235–237.

6. Zundler, S.; Neurath, M.F. Interleukin-12: Functional activities and implications for disease. *Cytokine Growth Factor Rev.* **2015**, *26*, 559–568. [CrossRef] [PubMed]

7. Guo, N.; Wang, W.Q.; Gong, X.J.; Gao, L.; Yang, L.R.; Yu, W.N.; Shen, H.Y.; Wan, L.Q.; Jia, X.F.; Wang, Y.S.; et al. Study of recombinant human interleukin-12 for treatment of complications after radiotherapy for tumor patients. *World J. Clin. Oncol.* **2017**, *8*, 158–167. [CrossRef]

8. Berraondo, P.; Etxeberria, I.; Ponz-Sarvise, M.; Melero, I. Revisiting interleukin-12 as a caner immunotherapy agent. *Clin. Cancer Res.* **2018**, *24*, 2716–2718. [CrossRef]

9. Tugues, S.; Burkhard, S.H.; Ohs, I.; Vrohlings, M.; Nussbaum, K.; Vom Berg, J.; Kulig, P.; Becher, B. New insights into IL-12-mediated tumor suppression. *Cell Death Differ.* **2015**, *22*, 237–246. [CrossRef] [PubMed]

10. Jayanthi, S.; Koppolu, B.P.; Smith, S.G.; Jalah, R.; Bear, J.; Rosati, M.; Pavlakis, G.N.; Felber, B.K.; Zaharoff, D.A.; Kumar, T.K. Efficient production and purification of recombinant human interleukin-12 (IL-12) overexpressed in mammalian cells without affinity tag. *Protein Expr. Purif.* **2014**, *102*, 76–84. [CrossRef] [PubMed]

11. Wang, P.; Li, X.; Wang, J.; Gao, D.; Li, Y.; Li, H.; Chu, Y.; Zhang, Z.; Liu, H.; Jiang, G.; et al. Re-designing Interleukin-12 to enhance its safety and potential as an anti-tumor immunotherapeutic agent. *Nat. Commun.* **2017**, *8*, 1395. [CrossRef] [PubMed]

12. Lo, K.M.; Lan, Y.; Lauder, S.; Zhang, J.; Brunkhorst, B.; Qin, G.; Verma, R.; Courtenay-Luck, N.; Gillies, S.D. huBC1-IL12, an immunocytokine which targets EDB-containing oncofetal fibronectin in tumors and tumor vasculature, shows potent anti-tumor activity in human tumor models. *Cancer Immunol. Immunother.* **2007**, *56*, 447–457. [CrossRef] [PubMed]

13. Sommavilla, R.; Pasche, N.; Trachsel, E.; Giovannoni, L.; Roesli, C.; Villa, A.; Neri, D.; Kaspar, M. Expression, engineering and characterization of the tumor-targeting heterodimeric immunocytokine F8-IL12. *Protein Eng. Des. Sel.* **2010**, *23*, 653–661. [CrossRef] [PubMed]

14. Tsai, J.L.; Priya, T.A.; Hu, K.Y.; Yan, H.Y.; Shen, S.T.; Song, Y.L. Grouper interleukin-12, linked by an ancient disulfide-bond architecture, exhibits cytokine and chemokine activities. *Fish Shellfish Immunol.* **2014**, *36*, 27–37. [CrossRef]

15. Yoon, C.; Johnston, S.C.; Tang, J.; Stahl, M.; Tobin, J.F.; Somers, W.S. Charged residues dominate a unique interlocking topography in the heterodimeric cytokine interleukin-12. *Embo J.* **2000**, *19*, 3530–3541. [CrossRef]

16. Oliveira, C.; Domingues, L. Guidelines to reach high-quality purified recombinant proteins. *Appl. Microbiol. Biotechnol.* **2018**, *102*, 81–92. [CrossRef] [PubMed]

17. Rogers, R.S.; Nightlinger, N.S.; Livingston, B.; Campbell, P.; Bailey, R.; Balland, A. Multi-attribute method for characterization, quality control testing and disposition of biologics. *Mabs-Austin* **2015**, *7*, 881–890. [CrossRef]

18. Rogers, R.S.; Abernathy, M.; Richardson, D.D.; Rouse, J.C.; Sperry, J.B. A view on the importance of "multi-attribute method" for measuring purity of biopharmaceuticals and improving overall control strategy. *AAPS J.* **2017**, *20*, 7. [CrossRef]

19. Reynolds, J.A.; Tanford, C. Binding of Dodecyl Sulfate to Proteins at High Binding Ratios. Possible Implications for the State of Proteins in Biological Membranes. *Proc. Natl. Acad. Sci. USA* **1970**, *66*, 1002–1007. [CrossRef]

20. Madej, T.; Lanczycki, C.J.; Zhang, D.; Thiessen, P.A.; Geer, R.C.; Marchler-Bauer, A.; Bryant, S.H. MMDB and VAST+: Tracking structural similarities between macromolecular complexes. *Nucleic Acids Res.* **2014**, *42*, D297–D303. [CrossRef]

21. Hosaka, M.; Nagahama, M.; Kim, W.S.; Watanabe, T.; Hatsuzawa, K.; Ikemizu, J.; Murakami, K.; Nakayama, K. Arg-X-Lys/Arg-Arg motif as a signal for precursor cleavage catalyzed by furin within the constitutive secretory pathway. *J. Biol. Chem.* **1991**, *266*, 12127–12130. [PubMed]

22. Matos, J.L.; Fiori, C.S.; Silva-Filho, M.C.; Moura, D.S. A conserved dibasic site is essential for correct processing of the peptide hormone AtRALF1 in Arabidopsis thaliana. *FEBS Lett.* **2008**, *582*, 3343–3347. [CrossRef] [PubMed]

23. Yadav, K.; Pathak, D.C.; Saikia, D.P.; Debnath, A.; Ramakrishnan, S.; Dey, S.; Chellappa, M.M. Generation and evaluation of a recombinant Newcastle disease virus strain R2B with an altered fusion protein cleavage site as a vaccine candidate. *Microb. Pathog.* **2018**, *118*, 230–237. [CrossRef] [PubMed]

Sample Availability: Samples of rhIL-12 are available from the authors.

molecules

MDPI

Article

A Simple Method for On-Gel Detection of Myrosinase Activity

Sándor Gonda [1,*], Zsolt Szűcs [1], Tamás Plaszkó [1], Zoltán Cziáky [2], Attila Kiss-Szikszai [3], Gábor Vasas [1] and Márta M-Hamvas [1]

[1]   Department of Botany, Division of Pharmacognosy, University of Debrecen, Egyetem tér 1, H-4010 Debrecen, Hungary; vapenda@gmail.com (Z.S.); plaszkotomi@gmail.com (T.P.); vasas.gabor@science.unideb.hu (G.V.); hamvas.marta@science.unideb.hu (M.M.-H.)

[2]   Agricultural and Molecular Research and Service Institute, University of Nyíregyháza, Sóstói str. 31/b, H-4400 Nyíregyháza, Hungary; cziaky.zoltan@nye.hu

[3]   Department of Organic Chemistry, University of Debrecen, Egyetem tér 1, H-4010 Debrecen, Hungary; attyska@gmail.com

*   Correspondence: gondasandor@gmail.com or gonda.sandor@science.unideb.hu; Tel.: +36-52-512-900/62634; Fax: +36-52-512-943

Received: 5 July 2018; Accepted: 28 August 2018; Published: 31 August 2018

Abstract: Myrosinase is an enzyme present in many functional foods and spices, particularly in Cruciferous vegetables. It hydrolyses glucosinolates which thereafter rearrange into bioactive volatile constituents (isothiocyanates, nitriles). We aimed to develop a simple reversible method for on-gel detection of myrosinase. Reagent composition and application parameters for native PAGE and SDS-PAGE gels were optimized. The proposed method was successfully applied to detect myrosinase (or sulfatase) on-gel: the detection solution contains methyl red which gives intensive red bands where the HSO_4^- is enzymatically released from the glucosinolates. Subsequently, myrosinase was successfully distinguished from sulfatase by incubating gel bands in a derivatization solution and examination by LC-ESI-MS: myrosinase produced allyl isothiocyanate (detected in conjugate form) while desulfo-sinigrin was released by sulfatase, as expected. After separation of 80 µg protein of crude extracts of Cruciferous vegetables, intensive color develops within 10 min. On-gel detection was found to be linear between 0.031–0.25 U (pure *Sinapis alba* myrosinase, $R^2 = 0.997$). The method was successfully applied to detection of myrosinase isoenzymes from horseradish, Cruciferous vegetables and endophytic fungi of horseradish as well. The method was shown to be very simple, rapid and efficient. It enables detection and partial characterization of glucosinolate decomposing enzymes without protein purification.

Keywords: myrosinase; thioglucosidase; sulfatase; on-gel detection; desulfo-sinigrin; LC-ESI-MS

1. Introduction

The glucosinolate-myrosinase-isothiocyanate system is a widely distributed chemical defense system of the Brassicales [1]. As the volatile isothiocyanates are highly bioactive molecules that are at the same time beneficial to human consumers, the system is of high scientific and industrial interest [2].

The plants biosynthesize the glucosinolate precursors, which come in contact with their activation enzyme myrosinase under some circumstances, usually when tissue damage occurs [1]. The reaction catalyzed by myrosinase (EC 3.2.1.147) is a thioglucoside hydrolysis which results in an unstable thiohydroximate that subsequently undergoes spontaneous rearrangement (Figure 1). The default products are isothiocyanates, or, in vivo in the presence of so called specifier proteins, other less toxic volatiles can be formed. The activity itself is shown to be present in various Brassicaceae plants [3,4], microorganisms [5–7], and organisms associated with such plants, like endophytes from horseradish roots [8] or some insects feeding on host plants with such metabolites [9–11]. A simple plant can contain various myrosinase isoenzymes, as shown in *Arabidopsis thaliana* [12] and other Brassicaceae plants [13–16].

Full characterization of enzymes requires purification during which activity is usually monitored by a routine, specific assay. Purification of enzymes with this activity was successful from several sources, including various Brassicaceae plants [4,13,17,18], microorganisms [6,7] or insects [11,19]. Myrosinase assays usually detect either the decomposition of the substrate (glucosinolates), or release of one of the product compounds. The possibilities include direct detection of volatile products by GC-MS [20,21], measurement of the released acid (pH stat assay) [22], a decrease of concentration of the glucosinolate substrate via spectrophotometry or chromatographic techniques [23,24], or derivatization of the released glucose (Figure 1) via coupled enzyme reactions [25]. The detection of the enzyme by immunological methods [26–28] is also a viable option.

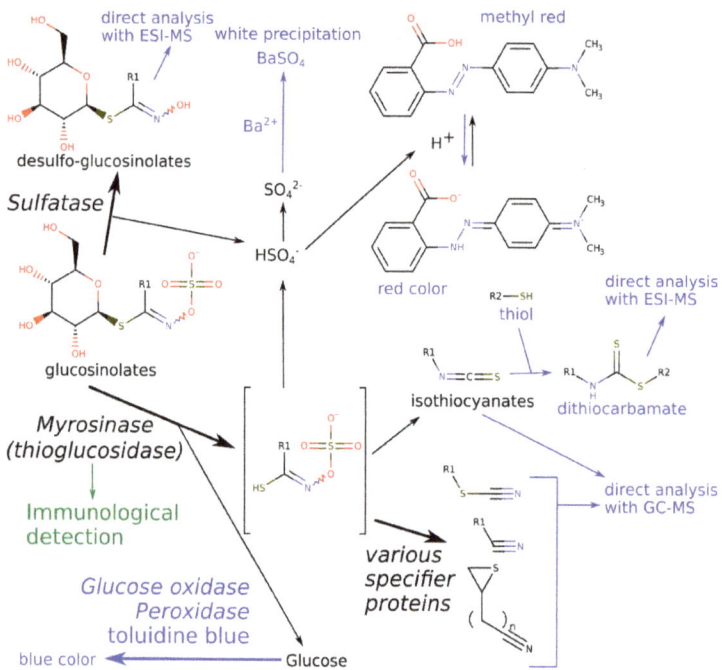

Figure 1. The main breakdown scheme of glucosinolates with possible detection methods of the products. Enzymatic reactions are shown with bold arrows, enzymes have italic font. The reactions that occur in vivo, are shown in black. Detection methods are either blue (those which require an operating myrosinase) or green (methods reliant only on similarity of sequences). The default rearrangement products of the thiohydroximates are isothiocyanates, alternative reaction products (from top to bottom) are nitriles, thiocyanates and epithionitriles. In our model, R1 = allyl- (sinigrin is converted to allyl isothiocyanate), R2 = 2-aminoethyl- (cysteamine).

Detection by immunological methods are very specific, and monoclonal anti-myrosinase antibodies, such as 3D7 are available for research [13,27,28]. However, while they can also detect myrosinases in inactive (denaturated) form, they are unable to detect proteins that have myrosinase activity, but are structurally unrelated to that used for antibody production. As they have evolved independently, myrosinases from different sources can vary considerably, which is perhaps highlighted by the fact that ascorbic acid inhibits the myrosinases of a cabbage aphid [11], marginally activates the myrosinase of *Citrobacter* [6], and activates plant myrosinases to a high extent [13]. If a myrosinase is to be detected that has unrelated peptide sequence to that of the known ones, insufficient binding of the primary antibody is likely.

Unfortunately, the above activity-based methods interfere with later protein purification steps, or suffer from other limitations. On-gel detection methods of myrosinases (thioglucosidases) include that of [29–31] who used Ba^{2+} to form a whitish precipitate from the released SO_4^{2-}. This approach also worked when screening fungal isolates for such an activity [32]. However, there are also limitations. Ba^{2+} forms white precipitates with a variety of anions (phosphate, carbonate, citrate), detection and documentation of the pale white color in the transparent gels can be a challenge, especially when low amounts of myrosinase is present. Perhaps therefore, there are no exact sensitivity data presented for on-gel usage in the literature. Ba^{2+} can also strongly bind to some proteins (see e.g., [33]). Though no data is available on this phenomenon with special respect to myrosinases, it might interfere with subsequent purification and characterization. Another disadvantage is that Ba^{2+} is highly toxic and requires special disposal. Another approach to detect myrosinase activity was the method of [34] who used starch gels, and used the glucose-oxidase-peroxidase-*o*-toluidine mixture, which results in blue colors if free glucose is present. This method has the advantage that it is more specific for myrosinase (sulfatase does not liberate any glucose, Figure 1), but the obtained gel sample is not suitable for later purification or characterization because of the added enzymes.

Seeing the limitations of the above, we aimed to develop a sensitive, straightforward on-gel assay for myrosinase that can also be used to detect bands on native PAGE gels or SDS-PAGE gels, after a simple washout protocol. Also, the development of an LC-ESI-MS method capable of distinguishing myrosinase from sulfatase was aimed.

2. Results and Discussion

2.1. On-Gel Detection of Glucosinolate Decomposition

Our approach was to detect the release of H^+, a side product of the myrosinase catalyzed glucosinolate hydrolysis (Figure 1). The detection was planned to be accomplished using a pH indicator in a weakly buffered reaction mixture. As myrosinase operates over a wide pH range [13], the produced acidification, detected by the pH indicator, does not inactivate the enzyme.

Preliminary tests to choose the proper pH indicator were done in test tubes, using crude extracts of horseradish roots containing high amounts of myrosinase. Congo red, bromocresol green and methyl red were selected for the test, as they show color transition within the range pH 4–6. Methyl red was chosen for further work as it provided the most spectacular color change during the in-vial assay, its color transition from yellow to red can be observed in the pH range 6–4.4. After addition of 10 μL myrosinase containing horseradish crude extract (typical protein content 45 μg) to 90 μL of the unbuffered detection reagent, the mixture developed intensive reddish color usually within a few minutes.

The least buffering capacity that still resulted in a stable solution (i.e., no spontaneous acidification and color change within 24 h) was found to be 1 mM phosphate, pH 7.5, methyl red concentration was 100 μg mL^{-1}. This was supplemented with the amount of sinigrin (6 mM) and ascorbic acid (1 mM) usually used in on-gel detection assays [31]. Hence, the final composition of the detection reagent was 6 mM sinigrin, 1 mM ascorbate, 1 mM Na_2HPO_4, pH 7.5, 100 μg mL^{-1} methyl red.

The detection reagent was successfully used "as is" for on-gel detection. After washing of native gels containing separated proteins of horseradish crude extracts, the myrosinase containing bands were successfully detected using the proposed detection reagent. Many enzymes can release acid, but the reaction conditions made the assay specific to glucosinolate decomposition: the proposed detection reagent does not produce color change in the absence of sinigrin as shown in a PAGE of horseradish crude extracts (Figure S1).

The reaction was also positive with purified myrosinase. The bands from purified myrosinase from *Sinapis alba* seeds (Sigma Aldrich, St. Louis, MO, USA) at 0.031–0.25 U, developed colors within 8 min (Figure 2, Figures S2c and S3). The gel in Figure 2 was photographed at 4, 6 and 8 min and evaluated by CP Atlas 2.0 gel image processing software (green channel). At 0.125 and 0.25 U, linear relationship was found between the signal and the incubation time ($R^2 \geq 0.996$). At 4 min, $R^2 = 0.997$ signal—activity linearity was obtained in the range 0.031–0.25 U (also see Table S1). This means that besides qualitative detection, approximate activity data can be obtained using the proposed method, within the given activity range. It is worth to note, that in case of extremely low activities, it was possible to left the gel covered for hours to detect minute amounts of myrosinase: no spontaneous acidification (red background increase) was observed in such gels, supporting the stability of the mixture in the absence of enzymes. The same amount of myrosinase did not result any white bands of $BaSO_4$ precipitation after the attempt to detect with the detection reagent of [31].

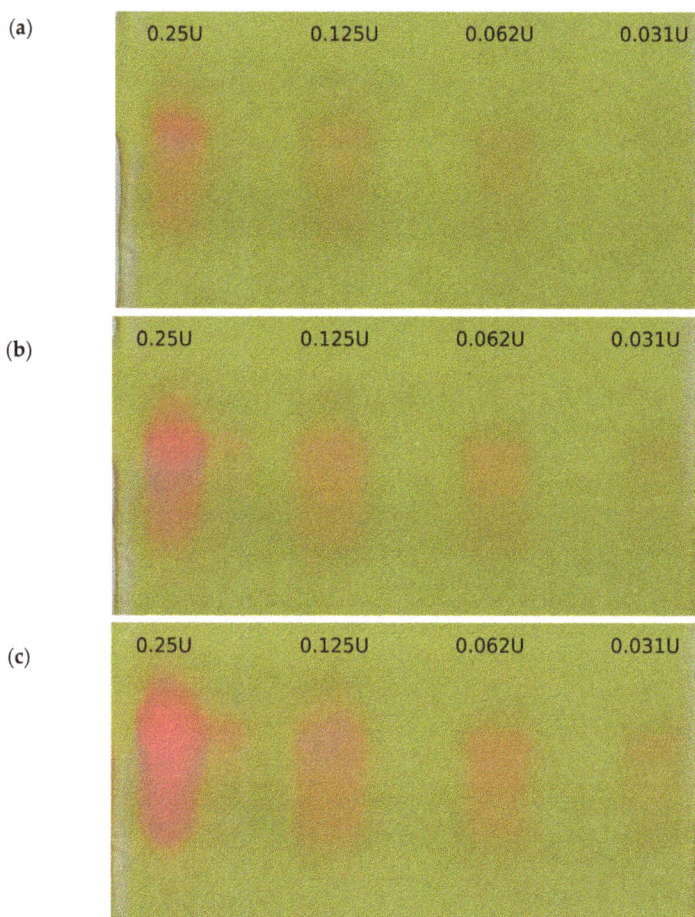

Figure 2. Short-term time course of the detectable on-gel signal—a serial dilution of myrosinase standard was detected with the proposed detection reagent containing sinigrin, 6 mM; ascorbic acid, 1 mM; Na_2HPO_4, 1 mM; pH 7.5; methyl red, 100 μg mL^{-1}. 0.031–0.25 enzyme units (U) of *Sinapis alba* thioglucosidase (myrosinase) standard) was separated on 7.5% native PAGE. Subplots: (**a**) 4 min, (**b**) 6 min, (**c**) 8 min.

The *S. alba* thioglucosidase as well as other plant myrosinase enzymes retained their activity after separation on SDS-PAGE gels and washout of SDS (Figure 3b, Figures S2b,c and S3, Table S2). At 4 min, $R^2 = 0.9899$ signal—activity was also observed (Table S2). The proposed washout procedure ensures elimination of the high amount of buffer and SDS that is typical during gel electrophoresis.

If desired, a higher sensitivity can be reached (at the cost of lower stability) by using a detection reagent with a slightly lower pH, this results in earlier color development (Figure S4).

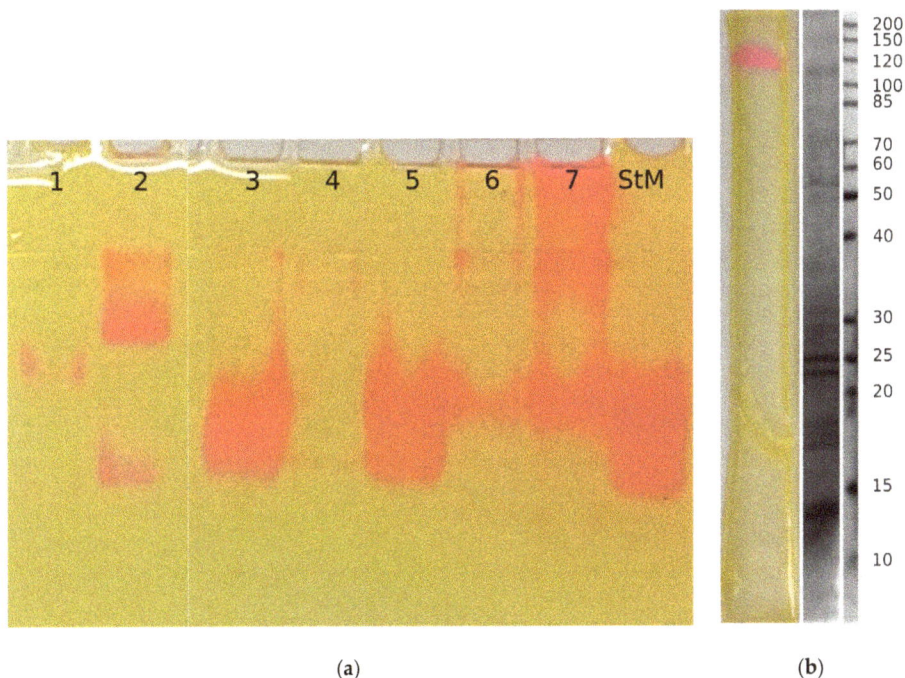

(a)   (b)

Figure 3. On-gel detection reaction of myrosinases from crude extracts of different species of Brassicaceae. Subplots: (**a**) Crude extracts with 80 µg protein content were loaded on 7.5% native PAGE. Samples: 1: *Brassica oleracea var. gemmifera* buds; 2: *Brassica oleracea var. italica* flowering heads, 3: rocket salad (*Eruca sativa*) whole seedlings, 4: *Brassica oleracea var. botrytis* flowering heads, 5: *Sinapis alba* whole seedlings, 6–7: *Armoracia rusticana* roots from two different sources, StM: *Sinapis alba* myrosinase standard. (**b**) 10% SDS-PAGE of horseradish root (*Armoracia rusticana*, 80 µg total protein) crude extracts. Myrosinase activity was detected by the proposed detection reagent after wash-out of SDS (**left**). The protein pattern of the horseradish root crude extract (**center**), and the molecular weight marker (Page RulerTM Unstained Protein Ladder, Thermo Scientific, **right**) were stained with Coomassie-Brillant Blue.

2.2. LC-ESI-MS of Products of Separated Sulfatase and Myrosinase Enzymes

As sulfatase also releases H^+ and SO_4^{2-} from sinigrin (Figure 1), a distinction has to be made whenever there is a possibility that sulfatase is present instead of myrosinase. This was carried out by LC-ESI-MS. We expected that incubation of sinigrin with myrosinase results in the release of allyl isothiocyanate (AITC) while sulfatase produces desulfosinigrin (Figure 1). LC-ESI-MS has the ability to detect desulfosinigrin with high sensitivity and specificity, while ITCs are usually detected by GC-MS, however, we wanted a single distinguishing measurement. Therefore, given our experience with ITC derivatization with thiols [24], we tested several thiols for ITC derivatization and LC-MS detection.

Several thiols with various side-chains that can be expected to ionize well under ESI-MS conditions were purchased and tested to react with 1 µg mL^{-1} allyl isothiocyanate in 50 mM NH$_4$OAc buffer (pH 9.0). The amount of ITC-thiol adduct (dithiocarbamate) produced is a function of pH [24], which is controlled by addition of the buffering agent at 40-fold excess compared to the reagent.

After testing several derivatization reagents, cysteamine was found to form the most sensitively detectable ITC derivates (Figure 1). The calibration curve in LC-ESI-MS was linear in the ranges 100–10^5 ng mL^{-1} for the cysteamine derivate (AITC equivalent, injecting 1 µL of derivatized sample). The LOD for cysteamine–AITC adduct was as low as 1 ng mL^{-1}. As compared to other LC-ESI-MS methods, this LOD is in the range of the most sensitive ITC determination methods [35,36]. MS/MS characterization of the cysteamine adduct (m/z 177.051) showed a major characteristic fragment 160.0248 ([M − NH$_3$ + H]$^+$, calcd. 160.0249, difference 0.6 ppm) as well as less intensive fragments m/z 101.0423 and 73.0112.

After optimization of the derivatization method and the LC-ESI-MS parameters, sulfatase standard (*Helix pomatia*), myrosinase standard (*Sinapis alba*) and horseradish crude extract were separated on slab gels and detected using the proposed detection reagent. The *Sinapis alba* myrosinase and horseradish crude extract yielded two and three bands, respectively, while sulfatase standard contained only one. Thereafter, the active bands were cut out and incubated in a derivatization solution containing 0.24 mM sinigrin, 100 mM NaH$_2$PO$_4$ buffer (final pH 7.5), 1 mM ascorbic acid and 46.1 mM cysteamine. Our previous study has shown that small aliphatic thiols are compatible with the myrosinase reaction, no enzyme inhibition was observed with mercaptoacetic acid in the reaction medium [24].

As expected, reaction with the band containing sulfatase resulted in production of desulfo-sinigrin, while reaction with all of the myrosinase bands (three from horseradish or two from *Sinapis alba*) resulted in allyl isothiocyanate which subsequently spontaneously conjugated with cysteamine (Figure 1 and Figure S5). Desulfosinigrin was identified using literature data [37]. The monoisotopic exact mass of the [M + H]$^+$ obtained was 280.0846 which matched the calculated value of 280.0849 (difference: −1.1 ppm). In MS/MS, the thioglycoside loses the glucose moiety, leading to a characteristic fragment C$_4$H$_8$NOS at m/z 118.0323 (calculated 118.0321, difference: −1.7 ppm). In case of the allyl-isothiocyanate adduct, the previously detected dithiocarbamate product was formed and detected by LC-ESI-MS at m/z 177.051. Hence, sulfatase and myrosinase were successfully distinguished using LC-ESI-MS without addition of any substance or enzyme that might hinder possible later protein purification or characterization steps.

2.3. On-Gel Detection after Separation of Crude Extracts

The main applications of the given detection method is to aid myrosinase purifications, and to study glucosinolate decomposing enzyme pattern of different organisms, which was achieved successfully using the proposed procedure: a series of crude extracts from different Brassicaceae plants were separated and their glucosinolate decomposing enzymes detected (Figure 3a). It is apparent, that most vegetables contain different complexes or isoenzymes that show the activity of interest. In case of the three isoenzymes from horseradish root crude extract, all three isoenzymes were proven to be myrosinases by detecting their reaction product allyl-isothiocyanate by LC-ESI-MS. As sulfatase activity is not described from Brassicaceae plants so far, the active bands of other vegetables most likely myrosinase isoenzymes or different complexes thereof. Perhaps because of the heavy glycosylation of these enzymes, separation efficacy in native PAGE and SDS-PAGE were similar (Figure S3a,b). In horseradish samples separated on SDS-PAGE, the MW of the myrosinase band approximately matches that published by [13] (Figure 3b).

Successful detection of glucosinolate-decomposing enzymes was also achieved from two of the endophytic fungi of horseradish, *Fusarium oxysporum* and *Macrophomina phaseolina*, which were recently shown to possess decomposing activity towards plant glucosinolates [8] (Figure S6). As allyl-isothiocyanate-GSH conjugate was detected from the spent medium, but desulfoglucosinolates were not [8], these organisms also likely possess myrosinase activity. The crude extract of the fungi (prepared with the same methodology as the crude plant extract) were shown to be orders of magnitude less active than that of the vegetables.

3. Materials and Methods

3.1. Chemicals

All reagents were of analytical purity. Allyl isothiocyanate and cysteamine were from Sigma-Aldrich (St. Louis, MO, USA). Congo red, bromocresol green and methyl red were from Reanal (Budapest, Hungary). Sinigrin was from Phytoplan (Heidelberg, Germany). Buffer components (disodium phosphate, ammonia, acetic acid, ammonium acetate) and ascorbic acid were from VWR (Debrecen, Hungary). As water, type I water (18.2 MΩ cm^{-1}) was used, which was produced by a Human Zeneer Power I water purification system (Human corporation, Seoul, Republic of Korea).

3.2. Pure Enzymes

Myrosinase (thioglucosidase) standard from *Sinapis alba* was from Sigma Aldrich, 1 U is equivalent to 1.0 μmole glucose min^{-1} from sinigrin at pH 6.0 at 25 °C. The used lot showed 260 U g^{-1} activity. Sulfatase standard from *Helix pomatia* was from Sigma Aldrich, 1 U was equivalent to hydrolysis of 1.0 μmole *p*-nitrocatechol sulfate h^{-1} at pH 5.0 at 37 °C (30 min assay). The enzyme used showed 30,320 U g^{-1} activity.

3.3. Organisms

Wild or commercial samples of the following plants were used: horseradish (*Armoracia rusticana*) root; *Armoracia macrocarpa* leaf; white mustard (*Sinapis alba*) seeds and whole seedlings; black mustard (*Sinapis nigrum*) seeds, black radish (*Raphanus sativus* var. *sativus*) root, rocket salad (*Eruca sativa*) leaf and seedlings, Brussels sprouts (*Brassica oleracea* var. *gemmifera*) buds, broccoli (*Brassica oleracea* var. *italica*) flowering heads and cauliflower (*Brassica oleracea* var. *botrytis*) flowering heads.

Selected horseradish endophytes from our previous study [8] (*Fusarium oxysporum*, *Macrophomina phaseolina*) were grown in inactivated horseradish extract until reaching stationary phase of growth in which decomposition of sinigrin usually initiates (5 and 7 days, respectively). The inactivated horseradish extract was prepared using boiling methanol, therefore it does not contain any active enzymes. It was standardized to contain 12 mM (5 mg mL^{-1}) sinigrin. The preparation of the inactivated horseradish extract and the inoculation parameters were the same as in our recent work [8].

3.4. Sample Preparation

The enzyme containing crude extracts were prepared as in our recent study [24]. Briefly, the raw plant material was mixed in a blender with cold 25 mM phosphate buffer (pH 6.5), centrifuged (13,000 rpm, 5 min), filtered on PES 0.22 μm membrane if necessary, and used directly for gel electrophoresis after protein content determination by a Bradford assay [38] as described earlier.

3.5. Testing of Detection Reagents

Congo red, bromocresol green or methyl red was dissolved in MeOH (1 mg mL^{-1}), and added to a mixture containing 6 mM sinigrin, 1 mM ascorbic acid, pH 8.0. To a 90 μL aliquot of the above mixture, 10 μL of horseradish crude extract (equivalent to about 45 μg protein) was added. The reaction mixture was subsequently incubated for about 15–30 min. As a positive control, 10 μL 0.1 M HCl was added to the mixture. To negative controls, 10 μL distilled water was added.

The final solution consisted of 1 mM Na$_2$HPO$_4$, 1mM ascorbic acid, 100 μg mL^{-1} methyl red (from a 1 mg mL^{-1} solution of methyl red in MeOH), 6 mM (2.5 mg mL^{-1}) sinigrin. The pH of the solution was adjusted to 8.0 with 0.1 M NaOH solution before adding methyl red and sinigrin. Addition of the residual components decreased the pH to a final value of 7.5.

3.6. Gel-Electrophoresis

Polyacrylamide slab gels (5.7% top/stacking and 10 or 7.5% bottom/resolving gels) with discontinuous buffer system were prepared according to [31,39] and used for on-gel detection of myrosinase. The optimal protein content of samples loaded onto gels were 20–200 µg depending on plant samples. After electrophoresis at 4 °C in dark, 20–22 mA/gel, gels were washed several times in purified/distilled water at room temperature to wash out the buffering electrolytes. For the successful reaction, the measured pH of the spent washing water was comparable to that of the washing distilled water (6.85–7.4). In our setup, samples were allowed to overrun for 35 min which made the separation of the more slowly migrating proteins better.

10 and 7.5% SDS-PAGE were used to investigate the ability of myrosinases for renaturation. Plant crude extracts were loaded together with molecular weight marker (PageRuler Unstained Protein Ladder 200–10 kDa, BLUeye prestained Protein Ladder 245–11 kDa, Sigma-Aldrich) to estimate molecular weight. After electrophoresis, renaturation of proteins was performed by washing out SDS in 20% (*v*/*v*) 2-propanol in water (2 × 15 min.). These steps were followed by washing with sterilized distilled water to obtain the optimal pH.

3.7. On-Gel Detection

For the reaction, 4 mL detection reagent was uniformly distributed on the surface of 90 × 120 mm gel on a glass or plastic foil. When samples with very different myrosinase activities are investigated, using of striped slab gels is recommended to avoid the cross-contamination between lanes. The typical reaction time of our plant samples was 10 min.

3.8. LC-ESI-MS

The UHPLC system (Ultimate 3000RS, Dionex, Sunnyvale, CA, USA) was coupled to a Thermo Q Exactive Orbitrap mass spectrometer (Thermo Fisher Scientific Inc., Waltham, MA, USA) equipped with an electrospray ionization source (ESI). The column was a Phenomenex Kinetex XB-C_{18} column (100 mm × 2.1 mm × 2.6 µm), oven temperature was maintained at 30 °C, flow rate was 250 µL min^{-1}. Eluent A was 100% water and eluent B was 100% acetonitrile. Both contained 0.1% formic acid as modifier. The following gradient elution program was used: 0 min, 95% A, 0–2 min, 95% A; 2–5 min, 75% A; 5–6 min, 40% A; 6–7 min, 0% A; 7–9 min, 0% A; 9–10 min, 95% A, 10–18 min, 95% A. 1 µL of the samples were injected in every run. The Q Exactive hybrid quadrupole-orbitrap mass spectrometer was operated in positive ion mode with the following parameters: capillary temperature 320 °C, spray voltage 4.0 kV, the resolution was set to 70,000. The mass range scanned was 150–1000 *m*/*z*. The maximum injection time was 100 ms. The resolution was set to 17,500 in the cases of MS^2 scans. The collision energy was 30 NCE. Sheath gas and aux gas flow rates were 32 and 7 arb, respectively.

3.9. Enzyme Reaction for Distinguishment of Myrosinase and Sulfatase

2.5 U of sulfatase and myrosinase standard as well as horseradish 80 µg protein was separated on a native gel. The gel pieces showing activity with the proposed reagent were washed with water and subsequently immersed in the derivatization solution, and incubated overnight at room temperature. The derivatization solution consisted of 3.56 mg mL^{-1} (46.1 mM) cysteamine, 100 mM NaH_2PO_4, 1 mM ascorbic acid and 240 mM (100 µg mL^{-1}) sinigrin. The pH of the solution was adjusted to 7.7 with 10M NaOH solution. The solution was diluted 25-fold with MeOH, centrifuged and 1 µL was injected to LC-ESI-MS as described above.

4. Conclusions

An inexpensive and simple method has been developed for on-gel detection of glucosinolate-decomposing enzymes, which is based on release of the $HSO_4{}^-$, i.e., the acidification of the reaction medium which is detected by methyl red pH indicator. When sufficient enzyme is present (which is the case when loading 80 μg protein for most tested Brassicaceae vegetables) the reaction is rapid (color develops within 10 min). Subsequent analysis by LC-MS allows one to distinguish myrosinase from sulfatase activity when necessary. The proposed detection method enables rapid detection of myrosinases and sulfatases from different sources after on-gel separation. We think the method can be of use for food chemistry research, functional characterization of the plant microbiome, chemical ecologists, and plant physiology studies as well.

Supplementary Materials: The supplementary are available on line.

Author Contributions: Study conception and design: S.G., acquisition of data: Z.S. (separation and on-gel detection, derivatization studies for LC-MS), T.P. (separation and on-gel detection), Z.C. (LC-MS), A.K.-S. (LC-MS), M.M.-H. (on-gel detection); interpretation of data: S.G., M.M.-H., Z.S.; drafting of manuscript: S.G., M.M.-H., Z.S., critical revision: G.V. All authors read and approved the final manuscript.

Funding: The study was supported by the Hungarian Scientific Research Fund (OTKA) 124339 and 128021 grants which is greatly acknowledged, as well as the European Union and the European Social Fund through projects EFOP-3.6.1-16-2016-00022 and the through project GINOP-2.3.2-15-2016-00008.

Conflicts of Interest: The authors declare no conflict of interest.

References

1. Kissen, R.; Rossiter, J.T.; Bones, A.M. The 'mustard oil bomb': Not so easy to assemble?! Localization, expression and distribution of the components of the myrosinase enzyme system. *Phytochem. Rev.* **2009**, *8*, 69–86. [CrossRef]

2. Nguyen, N.M.; Gonda, S.; Vasas, G. A Review on the Phytochemical Composition and Potential Medicinal Uses of Horseradish (*Armoracia rusticana*) Root. *Food Rev. Int.* **2013**, *29*, 1–15. [CrossRef]

3. Ludikhuyze, L.; Rodrigo, L.; Hendrickx, M. The Activity of Myrosinase from Broccoli (Brassica oleracea L. cv. Italica): Influence of Intrinsic and Extrinsic Factors. *J. Food Prot.* **2000**, *63*, 400–403. [CrossRef] [PubMed]

4. Bernardi, R.; Finiguerra, M.G.; Rossi, A.A.; Palmieri, S. Isolation and Biochemical Characterization of a Basic Myrosinase from Ripe Crambe abyssinica Seeds, Highly Specific for epi-Progoitrin. *J. Agric. Food Chem.* **2003**, *51*, 2737–2744. [CrossRef] [PubMed]

5. Smits, J.P.; Knol, W.; Bol, J. Glucosinolate degradation by Aspergillus clavatus and Fusarium oxysporum in liquid and solid-state fermentation. *Appl. Microbiol. Biotechnol.* **1993**, *38*, 696–701. [CrossRef]

6. Albaser, A.; Kazana, E.; Bennett, M.H.; Cebeci, F.; Luang-In, V.; Spanu, P.D.; Rossiter, J.T. Discovery of a Bacterial Glycoside Hydrolase Family 3 (GH3) β-Glucosidase with Myrosinase Activity from a Citrobacter Strain Isolated from Soil. *J. Agric. Food Chem.* **2016**, *64*, 1520–1527. [CrossRef] [PubMed]

7. Tani, N.; Ohtsuru, M.; Hata, T. Purification and general characteristics of bacterial myrosinase produced by Enterobacter cloacae. *Agric. Biol. Chem.* **1974**, *38*, 1623–1630. [CrossRef]

8. Szűcs, Z.; Plaszkó, T.; Cziáky, Z.; Kiss-Szikszai, A.; Emri, T.; Bertóti, R.; Sinka, L.T.; Vasas, G.; Gonda, S. Endophytic fungi from the roots of horseradish (*Armoracia rusticana*) and their interactions with the defensive metabolites of the glucosinolate-myrosinase-isothiocyanate system. *BMC Plant Biol.* **2018**, *18*, 85. [CrossRef] [PubMed]

9. Beran, F.; Pauchet, Y.; Kunert, G.; Reichelt, M.; Wielsch, N.; Vogel, H.; Reinecke, A.; Svatoš, A.; Mewis, I.; Schmid, D.; et al. Phyllotreta striolata flea beetles use host plant defense compounds to create their own glucosinolate-myrosinase system. *Proc. Natl. Acad. Sci. USA* **2014**, *111*, 7349–7354. [CrossRef] [PubMed]

10. Jones, A.M.E.; Bridges, M.; Bones, A.M.; Cole, R.; Rossiter, J.T. Purification and characterisation of a non-plant myrosinase from the cabbage aphid *Brevicoryne brassicae* (L.). *Insect Biochem. Mol. Biol.* **2001**, *31*, 1–5. [CrossRef]

11. Pontoppidan, B.; Ekbom, B.; Eriksson, S.; Meijer, J. Purification and characterization of myrosinase from the cabbage aphid (Brevicoryne brassicae), a brassica herbivore. *Eur. J. Biochem.* **2001**, *268*, 1041–1048. [CrossRef] [PubMed]

12. Barth, C.; Jander, G. Arabidopsis myrosinases TGG1 and TGG2 have redundant function in glucosinolate breakdown and insect defense. *Plant J.* **2006**, *46*, 549–562. [CrossRef] [PubMed]

13. Li, X.; Kushad, M.M. Purification and characterization of myrosinase from horseradish (*Armoracia rusticana*) roots. *Plant Physiol. Biochem.* **2005**, *43*, 503–511. [CrossRef] [PubMed]

14. Eriksson, S.; Ek, B.; Xue, J.; Rask, L.; Meijer, J. Identification and characterization of soluble and insoluble myrosinase isoenzymes in different organs of *Sinapis alba*. *Physiol. Plant.* **2001**, *111*, 353–364. [CrossRef] [PubMed]

15. Lenman, M.; Falk, A.; Rodin, J.; Hoglund, A.S.; Ek, B.; Rask, L. Differential Expression of Myrosinase Gene Families. *Plant Physiol.* **1993**, *103*, 703–711. [CrossRef] [PubMed]

16. Loebers, A.; Müller-Uri, F.; Kreis, W. A young root-specific gene (ArMY2) from horseradish encoding a MYR II myrosinase with kinetic preference for the root-specific glucosinolate gluconasturtiin. *Phytochemistry* **2014**, *99*, 26–35. [CrossRef] [PubMed]

17. Durham, P.L.; Poulton, J.E. Enzymic Properties of Purified Myrosinase from Lepidium sativum Seedlings. *Z. Naturforsch. C* **2014**, *45*, 173–178. [CrossRef]

18. Palmieri, S.; Iori, R.; Leoni, O. Myrosinase from *Sinapis alba* L.: A new method of purification for glucosinolate analyses. *J. Agric. Food Chem.* **1986**, *34*, 138–140. [CrossRef]

19. Francis, F.; Lognay, G.; Wathelet, J.-P.; Haubruge, E. Characterisation of aphid myrosinase and degradation studies of glucosinolates. *Arch. Insect Biochem. Physiol.* **2002**, *50*, 173–182. [CrossRef] [PubMed]

20. Lambrix, V.; Reichelt, M.; Mitchell-Olds, T.; Kliebenstein, D.J.; Gershenzon, J. The Arabidopsis Epithiospecifier Protein Promotes the Hydrolysis of Glucosinolates to Nitriles and Influences Trichoplusia ni Herbivory. *Plant Cell* **2001**, *13*, 2793–2807. [CrossRef] [PubMed]

21. Wittstock, U.; Agerbirk, N.; Stauber, E.J.; Olsen, C.E.; Hippler, M.; Mitchell-Olds, T.; Gershenzon, J.; Vogel, H. Successful herbivore attack due to metabolic diversion of a plant chemical defense. *Proc. Natl. Acad. Sci. USA* **2004**, *101*, 4859–4864. [CrossRef] [PubMed]

22. Piekarska, A.; Kusznierewicz, B.; Meller, M.; Dziedziul, K.; Namiesnik, J.; Bartoszek, A. Myrosinase activity in different plant samples; optimisation of measurement conditions for spectrophotometric and pH-stat methods. *Ind. Crop. Prod.* **2013**, *50*, 58–67. [CrossRef]

23. Vastenhout, K.J.; Tornberg, R.H.; Johnson, A.L.; Amolins, M.W.; Mays, J.R. High-performance liquid chromatography-based method to evaluate kinetics of glucosinolate hydrolysis by *Sinapis alba* myrosinase. *Anal. Biochem.* **2014**, *465*, 105–113. [CrossRef] [PubMed]

24. Gonda, S.; Kiss-Szikszai, A.; Szűcs, Z.; Nguyen, N.M.; Vasas, G. Myrosinase Compatible Simultaneous Determination of Glucosinolates and Allyl Isothiocyanate by Capillary Electrophoresis Micellar Electrokinetic Chromatography (CE-MEKC). *Phytochem. Anal.* **2016**, *27*, 191–198. [CrossRef] [PubMed]

25. Wilkinson, A.; Rhodes, M.; Fenwick, G. Determination of Myrosinase (thioglucoside Glucohydrolase) Activity by a Spectrophotometric Coupled Enzyme Assay. *Anal. Biochem.* **1984**, *139*, 284–291. [CrossRef]

26. Bones, A.M.; Thangstad, O.P.; Haugen, O.A.; Espevik, T. Fate of Myrosin Cells: Characterization of Monoclonal Antibodies Against Myrosinase. *J. Exp. Bot.* **1991**, *42*, 1541–1550. [CrossRef]

27. Andréasson, E.; Jørgensen, L.B.; Höglund, A.-S.; Rask, L.; Meijer, J. Different Myrosinase and Idioblast Distribution in Arabidopsis and Brassica napus. *Plant Physiol.* **2001**, *127*, 1750–1763. [CrossRef] [PubMed]

28. Lenman, M.; Rödin, J.; Josefsson, L.-G.; Rask, L. Immunological characterization of rapeseed myrosinase. *Eur. J. Biochem.* **1990**, *194*, 747–753. [CrossRef] [PubMed]

29. MacGibbon, D.B.; Allison, R.M. A method for the separation and detection of plant glucosinolases (myrosinases). *Phytochemistry* **1970**, *9*, 541–544. [CrossRef]

30. Bones, A.; Slupphaug, G. Purification, Characterization and Partial Amino Acid Sequencing of β-thioglucosidase from *Brassica napus* L. *J. Plant Physiol.* **1989**, *134*, 722–729. [CrossRef]

31. Shikita, M.; Fahey, J.W.; Golden, T.R.; Holtzclaw, W.D.; Talalay, P. An unusual case of "uncompetitive activation" by ascorbic acid: Purification and kinetic properties of a myrosinase from Raphanus sativus seedlings. *Biochem. J.* **1999**, *341*, 725–732. [CrossRef] [PubMed]

32. Rakariyatham, N.; Butrindr, B.; Niamsup, H.; Shank, L. Screening of filamentous fungi for production of myrosinase. *Braz. J. Microbiol.* **2005**, *36*, 242–245. [CrossRef]

33. Sharma, A.; Yogavel, M.; Sharma, A. Utility of anion and cation combinations for phasing of protein structures. *J. Struct. Funct. Gen.* **2012**, *13*, 135–143. [CrossRef] [PubMed]

34. Vaughan, J.G.; Gordon, E.; Robinson, D. The identification of myrosinase after the electrophoresis of Brassica and Sinapis seed proteins. *Phytochemistry* **1968**, *7*, 1345–1348. [CrossRef]

35. Wang, H.; Lin, W.; Shen, G.; Khor, T.-O.; Nomeir, A.A.; Kong, A.-N. Development and Validation of an LC-MS-MS Method for the Simultaneous Determination of Sulforaphane and its Metabolites in Rat Plasma and its Application in Pharmacokinetic Studies. *J. Chromatogr. Sci.* **2011**, *49*, 801–806. [CrossRef] [PubMed]

36. Platz, S.; Kuehn, C.; Schiess, S.; Schreiner, M.; Mewis, I.; Kemper, M.; Pfeiffer, A.; Rohn, S. Determination of benzyl isothiocyanate metabolites in human plasma and urine by LC-ESI-MS/MS after ingestion of nasturtium (*Tropaeolum majus* L.). *Anal. Bioanal. Chem.* **2013**, *405*, 7427–7436. [CrossRef] [PubMed]

37. Kusznierewicz, B.; Iori, R.; Piekarska, A.; Namieśnik, J.; Bartoszek, A. Convenient identification of desulfoglucosinolates on the basis of mass spectra obtained during liquid chromatography–diode array–electrospray ionisation mass spectrometry analysis: Method verification for sprouts of different Brassicaceae species extracts. *J. Chromatogr. A* **2013**, *1278*, 108–115. [CrossRef] [PubMed]

38. Bradford, M.M. A rapid and sensitive method for the quantitation of microgram quantities of protein utilizing the principle of protein-dye binding. *Anal. Biochem.* **1976**, *72*, 248–254. [CrossRef]

39. Laemmli, U.K. Cleavage of Structural Proteins during the Assembly of the Head of Bacteriophage T4. *Nature* **1970**, *227*, 680–685. [CrossRef] [PubMed]

Sample Availability: Not available. Sampling of wild plants complied with local legislation and guidelines.

molecules

MDPI

Communication

First Look at the Venom of *Naja ashei*

Konrad Kamil Hus [1], Justyna Buczkowicz [1], Vladimír Petrilla [2,3], Monika Petrillová [4], Andrzej Łyskowski [1], Jaroslav Legáth [1,5] and Aleksandra Bocian [1,*]

[1] Department of Biotechnology and Bioinformatics, Faculty of Chemistry, Rzeszow University of Technology, Powstańców Warszawy 6, 35-959 Rzeszow, Poland; knr.hus@gmail.com (K.K.H.); czaporj@prz.edu.pl (J.B.); alyskowski@prz.edu.pl (A.Ł.); Jaroslav.Legath@uvlf.sk (J.L.)

[2] Department of Physiology, University of Veterinary Medicine and Pharmacy, Komenského 73, 041 81 Kosice, Slovakia; petrillav@gmail.com

[3] Zoological Department, Zoological Garden Košice, Široká 31, 040 06 Košice-Kavečany, Slovakia

[4] Department of General Education Subjects, University of Veterinary Medicine and Pharmacy, Komenského 73, 041 81 Kosice, Slovakia; monika.petrillova@uvlf.sk

[5] Department of Pharmacology and Toxicology, University of Veterinary Medicine and Pharmacy, Komenského 73, 041 81 Kosice, Slovakia

[*] Correspondence: bocian@prz.edu.pl; Tel.: +48-17-865-1287

Received: 31 January 2018; Accepted: 6 March 2018; Published: 8 March 2018

Abstract: *Naja ashei* is an African spitting cobra species closely related to *N. mossambica* and *N. nigricollis*. It is known that the venom of *N. ashei*, like that of other African spitting cobras, mainly has cytotoxic effects, however data about its specific protein composition are not yet available. Thus, an attempt was made to determine the venom proteome of *N. ashei* with the use of 2-D electrophoresis and MALDI ToF/ToF (Matrix-Assisted Laser Desorption/Ionization Time of Flight) mass spectrometry techniques. Our investigation revealed that the main components of analysed venom are 3FTxs (Three-Finger Toxins) and PLA$_2$s (Phospholipases A$_2$). Additionally the presence of cysteine-rich venom proteins, 5′-nucleotidase and metalloproteinases has also been confirmed. The most interesting fact derived from this study is that the venom of *N. ashei* includes proteins not described previously in other African spitting cobras—cobra venom factor and venom nerve growth factor. To our knowledge, there are currently no other reports concerning this venom composition and we believe that our results will significantly increase interest in research of this species.

Keywords: *Naja ashei*; venom composition; 2-D electrophoresis; proteomics

1. Introduction

Electrophoretic techniques have been extensively used during past years to analyze complex mixtures of peptides and proteins like snake venoms. Rapid development of chromatographic techniques coupled with mass spectrometry is considered as standard in modern proteomics, however two-dimensional electrophoresis still remains an important method in analysis of venom variation, post-translational modifications or whole proteome mapping.

The African spitting cobras are widely distributed throughout the dry, open areas of sub-Saharan region. They are present from Senegal in the west to Somalia in the east, and from southern Egypt in the north to South Africa. This group comprises several snake species, including *Naja nigricollis*, *N. katiensis* and *N. pallida*. In 2007, *N. ashei* became another representative of African spitting cobras as Wüster and Broadley have classified it as a separate species [1]. In general, *N. ashei* venom has similar properties to the venoms of other African spitting cobras. It can cause local tissue damage, i.e., oedema, blistering and necrosis of the skin and subcutaneous connective tissue [2–5]. In addition, the venom is often spat into the eyes causing ophthalmic lesions [6]. After snake attack, a rapid development of tissue necrosis is observed, and in cases when antivenom treatment is administered too late, local lesions often lead to chronic ulceration, osteomyelitis, arthrodesis, hypertrophic scars, keloid formation and, in some chronic cases, malignant transformation [5].

Proteomic analysis of the spitting cobra venom composition revealed that in majority it consists of three-finger toxin (3FTx) and cytotoxic phospholipase A_2 (PLA$_2$) molecules accounting, respectively, for 67–73% and 22–30% of the total venom proteins. The third largest group of proteins are the snake venom metalloproteinases (SVMPs) from PIII subfamily. There are also some other proteins that are less universal for all African spitting cobras, for instance: nucleotidases, cysteine-rich secretory proteins (CRISPs) or nawaprin [7].

So far, to our knowledge, no one has undertaken an analysis of the protein or peptide composition of *Naja ashei* venom. Therefore, in our study we present for the first time our initial venom composition results determined with the use of 2-D electrophoresis coupled with MALDI ToF/ToF mass spectrometry analysis.

2. Results

Distribution of spots on the obtained gels clearly indicated that the vast majority of proteins in the *Naja ashei* venom have a low molecular weight and clearly basic character (Figure 1). On the gel there were about 80 spots in the pH range of 3–10. The exact number was impossible to determine because of the smears, spot trains and inaccurate separation of the most abundant spots.

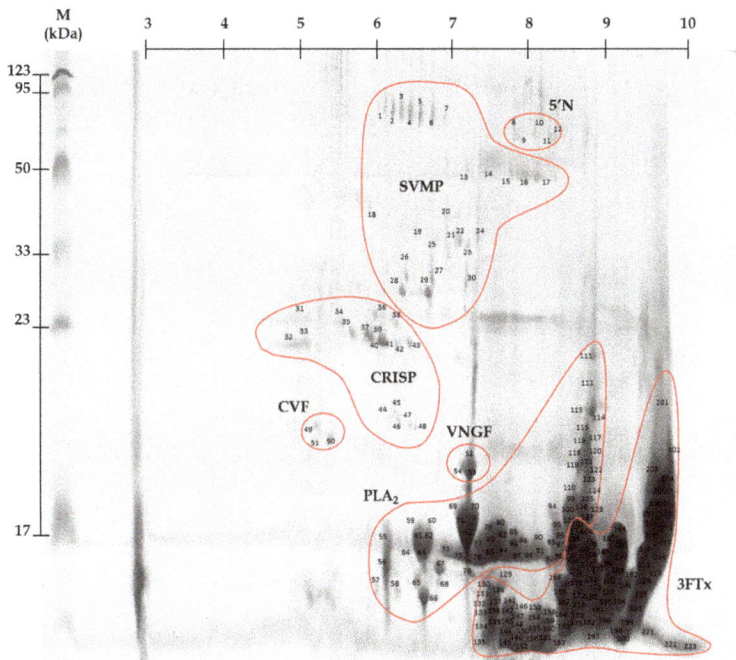

Figure 1. Representative 2-D protein map obtained from *Naja ashei* venom. **5′N**—Snake venom 5′-nucleotidase; **SVMPs**—Snake venom metalloproteinases; **CRISPs**—Cysteine-rich venom proteins; **CVF**—Cobra venom factor; **VNGF**—Venom nerve growth factor, **PLA$_2$s**—Phospholipases A$_2$; **3FTx**—Snake three-finger toxin family.

The results for protein identification using MALDI ToF/ToF mass spectrometry are summarized in Table 1. Identified proteins were grouped into seven major groups (Figure 2).

Table 1. Proteins identified in *Naja ashei* venom.

Gel Area[1]	Protein Name[2]	Protein Accession Code, Source Organism as Determined by Mascot and Spot Numbers[2]	Mass [kDa][3]	Score[4]	m/z[5]	MS/MS-Derived Sequence/Sequence Coverage[6]
SVMP	Zinc metalloproteinase-disintegrin-like cobrin	Q9PVK7 (*Naja kaouthia*) 20–24, 27	69 / 69	60 / 81	PMF / 1280.722	SC 9.5% / DPSYGMVEPGTK
	Zinc metalloproteinase-disintegrin-like atrase A	D5LMJ3 (*Naja atra*) 1–6	70 / 70 / 70	62 / 30 / 44	1087.732 / 1073.517 / 1497.840	EHQEYLLR / KGDDVSHCR / ERPQCILNKPSR
	Zinc metalloproteinase-disintegrin-like atragin	D3TTC2 (*Naja atra*) 14–17, 25, 26, 28–30	71 / 71 / 71	24 / 35 / 46	1140.664 / 1155.607 / 1476.894	DSCFTLNQR / CGDGMVCSNR / CPIMTNQCIALR
5'N	Snake venom 5'-nucleotidase	F8S0Z7 (*Crotalus adamanteus*) 8–12	57 / 63 / 65	48 / 62 / 32	1523.801 / 1389.797 / 1110.568	HGQGTGELLQVSGIK / LTILHNDVHAR / QAFEHSVHR
CRISP	Cysteine-rich venom protein annuliferin a (fragment)	PODL14 (*Naja annulifera*) 40, 44–48	3.6 / 3.6	68 / 96	1168.696 / 1195.609	NVDFNSESTR / EIVDLHNSLR
	Cysteine-rich venom protein natrin 1	Q7T1K6 (*Naja atra*) 32, 33, 37–39, 41–43	27 / 27	80 / 45	1553.910 / 1569.594	MEWYPEAASNAER / MEWYPEAASNAER
CVF	Cobra venom factor	Q91132 (*Naja kaouthia*) 49–51	185 / 185	37 / 58	1306.709 / 1337.885	GICVAEPYEIR / VNDDYLIWGSR
PLA₂	Acidic phospholipase A2 CM-I	P00602 (*Naja mossambica*) 55, 56, 59–63, 70, 71	14 / 14	60 / 110	PMF / 1769.783	32.2% / CCQVHDNCYGEAEK
	Basic phospholipase A2 1	P00603 (*Naja mossambica*) 57, 58, 65–68, 72–78	14 / 14 / 14	60 / 46 / 72	PMF / 987.512 / 1413.809	32.2% / GTPVDDLDR / LGCWPYLTLYK
	Basic phospholipase A2 CM-III	P00604 (*Naja mossambica*) 79–89, 91–93, 95, 101, 113–121, 123–127	14 / 14 / 14 / 14	90 / 99 / 79 / 193 / 28	PMF / 1374.965 / 1512.841 / 2157.377 / 1282.633	55.9% / YIDANYNINFK / CCQVHDNCYEK / CGAAVCNCDIVAANCFAGAR / CTVPSRSWWHFANYGCYCGR
VNGF	Venom nerve growth factor	P61898 (*Naja atra*) 53, 54	13 / 13 / 13	60 / 49 / 41	1127.664 / 1648.000 / 1415.821	NPNPEPSGCR / GNTVTVMENVNLDNK / CKNPNPEPSGCR
		Q90W38 (*Bothrops jararacussu*) 52	27 / 27 / 27	65 / 71 / 45	962.627 / 1363.885 / 1379.914	QYFFETK / ALTMEGNQASWR / ALTMEGNQASWR

Table 1. *Cont.*

Gel Area [1]	Protein Name [2]	Protein Accession Code, Source Organism as Determined by Mascot and Spot Numbers [2]	Mass [kDa] [3]	Score [4]	m/z [5]	MS/MS-Derived Sequence/Sequence Mass Coverage [6]
	Cytotoxin 1	P01467 (*Naja mossambica*) [C] 103–105	7 7	56 68	PMF 1302.807	45% CNQLIPPFWK
		P01468 (*Naja pallida*) [C] 139, 144–147, 166–192, 200, 204–220	7 7	78 50	PMF 1091.463	58.3% YMCCNTDK
	Cytotoxin 2	P01469 (*Naja mossambica*) [C] 193–196, 221–223	7 7	59 50	PMF 948.463	45% GCIDVCPK
3FTx	Cytotoxin 4	P01452 (*Naja mossambica*) [C] 106–109, 162, 180, 206	7	40	1060.609	YVCCSTDR
	Cytotoxin 5	P25517 (*Naja mossambica*) [C] 109, 142, 148–157, 159–163, 180, 206	7	39	1118.459	YECCDTDR
	Cytotoxin 11	P62390 (*Naja annulifera*) [C] 130, 136, 138–143, 154, 159, 164	7	52	1020.337	RGCAATCPK
	Muscarinic toxin-like protein 2	P82463 (*Naja kaouthia*) [M] 131	7	69	1319.692	GCAATCPIAENR

[1] Spot gel area name is the same as in Figures 1 and 2; [2] Protein name and database accession number of homologous proteins and organism from which protein identification originates. In the case of 3FTx cytotoxins: [C] cytotoxin activity, [M] muscarinic toxin-like activity (according to the UniProt database, www.uniprot.org). Spot numbers are related to Figure 1; [3] The mass of molecule as reported by Mascot (Boston, MA, USA); [4] Protein identification was performed using the Mascot search with probability based Mowse score. Ions score was $-10 \times \log(P)$, where P was the probability that the observed match was a random event; [5] Mass of precursor ion or information about PMF (Peptide Mass Fingerprinting) identification mode use; [6] Peptide sequence derived from LIFT analysis (Autoflex Speed, Bruker Daltonics, Billerica, MA, USA). Identification of proteins by MS/MS method was conducted by comparing obtained sequences with sequences from database. In the case of PMF identification: SC—amino acid sequence coverage for the identified proteins. Collection of annotated mass spectra is available as a Supplementary Material.

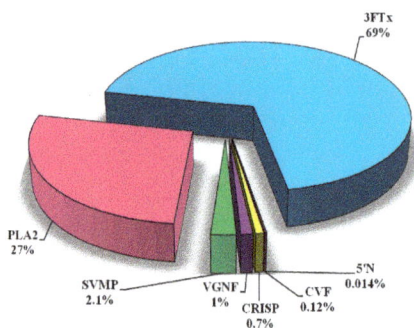

Figure 2. The percentage distribution of different protein groups in *Naja ashei* proteome calculated on the basis of %Vol of particular spots on gels. Abbreviations are the same as in Figure 1.

Using the %Vol of each spot on the gel, relative amounts of individual protein fractions in the venom of *N. ashei* were determined. Percentage distribution of protein groups is presented in Figure 2. According to this analysis, the most abundant proteins are cytotoxins belonging to snake three-finger toxins (almost 70%). The second highly abundant group are phospholipases A_2 (27%). The share of other groups of proteins: metalloproteinases, venom nerve growth factor, cysteine-rich venom proteins, cobra venom factor, snake venom 5′-nucleotidase does not exceed 5% of the total protein content (Figure 2).

3. Discussion

Proteomic analysis of the venoms of African species of spitting cobras has revealed similar properties and protein profiles [7]. We discovered seven groups of proteins, among which 3FTxs and phospholipases A_2 were the most abundant. The remaining five groups of proteins (SVMPs, CRISPs, venom nerve growth factor, cobra venom factor and 5′-nucleotidases) together constitute less than 5% of the total proteins of *Naja ashei* venom. For this group of Elapidae the predominant share of cytotoxic 3FTx and PLA$_2$ molecules is distinctive. However, the minor contribution of SVMPs, CRISPs, and endonucleases was also described [7]. On the basis of this composition, it is likely that the major cytotoxins and PLA$_2$s are responsible for the predominant myo- and cytotoxic effects induced by these venoms (i.e., dermonecrosis) [8].

A large number of three-finger toxins interfere with cholinergic transmission in the peripheral and central nervous system, thus, they are classified to the neurotoxin group [9]. However, a large number of the 3FTxs also exhibit general cytolytic properties (i.e., disruption of the membrane bilayer by forming pores in the cellular surface or penetrating into the biological membranes and triggering different biological phenomena and, therefore, they are also referred to as cytolysins or cytotoxins) [10–13]. The most interesting from a pharmacological point of view is the fact that cytotoxins possess significant and selective anticancer activity by inducing apoptosis or necrosis of tumor cells [14–20]. It makes this group a very interesting object of investigation, especially since the intact proteins from *Naja ashei* have never been examined.

The second most abundant protein group in *N. ashei* venom are phospholipases A_2 (Figure 2). In general, PLA$_2$s exhibit a wide variety of physiological and pathological effects. They undeniably play a role in the digestion of prey, but also exhibit a wide spectrum of pharmacological effects, such as neurotoxicity, cardiotoxicity, myotoxicity, and anticoagulant effect [21–27]. Interestingly, this group of proteins has also anticancer [28–32] and antimicrobial properties [33–36]. In *N. ashei* venom PLA$_2$s constitute 27% of all identified proteins, and this value is typical for all African spitting cobras [7].

The third group of proteins, distinctive for all African spitting cobras, are metalloproteinases. Their quantity in this group of snake venom ranges from 1.6 to 3.3% [7], and in *N. ashei* metalloproteinases share 2.1% of total venom proteins. All identified metalloproteinases belong to PIII family, and are zinc-dependent enzymes degrading plasma proteins and the extracellular matrix surrounding blood vessels, leading to local and systemic haemorrhage and coagulopathy [37–39]. PIII-SVMPs are present in venoms of all venomous snakes; however, their proportion in Elapid venom is much lower than in Viperid [40,41]. This fact determined that elapid SVMPs are much less understood, although it is believed that local tissue damage, haemorrhage, and complement depletion, reported after *N. nigricollis* bites, are caused by SVMP activity [2,4]. Low content of metalloproteinases in venom could indicate their minor role in the pathophysiology of envenoming. However, some studies reported that their high enzymatic activity can vastly contribute to the detrimental effects of venom [38,39,42].

CRISPs were also detected in *N. ashei* venom (Figure 1), however their content is definitely small (Figure 2). They are widely distributed among different snake venoms, and in our earlier studies, we have detected them in Viperidae venoms [41,43]. Intriguingly, this group of non-enzymatic proteins is not typical for all African spitting cobra species. Earlier works indicated their presence only in *N. nigricollis* and *N. katiensis* venoms [7]. 5′-Nucleotidase seems to be more universal for this group of snakes, because it was detected in all species except *N. nubiae* [7]. Enzymes from this group were detected in venoms of several species, always in small quantities [41,44,45].

It is very interesting that we were able to identify two proteins not detected before in African spitting cobras venom. They are: cobra venom factor, with 0.12% share, and venom nerve growth factor, with 1% share of total venom proteins. A negligible amount of these proteins in the venom indicates that their impact on the pathology of envenoming is low, but these proteins are extremely interesting from a pharmacological point of view. Cobra venom factor depletes complement C3 protein, and thus inhibits inflammatory and immune responses. This protein could be potentially used in several human diseases treatment, for instance: myocardial ischemia reperfusion injury, age-related macular degeneration, arthritis, paroxysmal nocturnal haemoglobinuria or lymphoma [46], and carcinoma [47]. In turn, VNGF is important for the growth, development, differentiation, and survival of neurons both in the peripheral and the central nervous systems [48], and additionally it inhibits metalloproteinase-disintegrin proteins [49]. It is known that nerve growth factors interact with some cancer cells [50,51], however the greatest hopes for their use lay in the treatment of neurodegenerative diseases [52–55].

This study shows that two-dimensional electrophoresis still can be used as an effective method for protein separation in analysis of snake venom proteome. Moreover, presented results clearly indicate that venom of *Naja ashei* is very similar to the closely related African spitting cobras. Nevertheless, the most interesting fact derived from this study is that the venom of *N. ashei* includes proteins not described so far in African spitting cobras. There are no other reports concerning this venom composition and we believe that our results will significantly increase interest in research of this species.

4. Materials and Methods

Pooled *Naja ashei* venom sample was obtained from two adult snakes (male and female), which were captured and officially imported from Kenya. Venom was extracted in the Pata breeding garden near Hlohovec (Slovakia), which had been designed for conservation of the reptiles' gene pool under the veterinary certificate No. CHEZ-TT-01. The breeding garden also serves as a quarantine station for imported animals and is an official importer of exotic animals from around the world, having the permission of the State Nature Protection of the Slovak Republic under the No. 03418/06, the trade with endangered species of wild fauna and flora and on amendments to certain laws under Law no. 237/2002 Z.z. After extraction, the venom was stored at −20 °C (transport temperature) and then moved to −80 °C for deep freezing.

The detailed procedure for proteomic analysis was described in our previous papers [41,43]. Protein concentration in crude venom was measured with 2-D Quant Kit (GE Healthcare, Little Chalfont, UK), using bovine serum albumin as a standard. The samples for isoelectrofocusing (IEF) were prepared by mixing 405 µg of proteins with standard thiourea rehydration solutions containing IPG buffers 3–10 pH range (GE Healthcare). Separation was conducted on 17 cm ReadyStrip IPG Strips with 3–10 pH gradient (Bio-Rad, Hercules, CA, USA). After IEF, the strips were incubated in equilibration buffers; one containing 1% DTT (for reduction); second containing 2.5% IAA (for alkylation). Prior to SDS-PAGE (Sodium dodecyl sulfate-polyacrylamide gel electrophoresis), gel strips were placed onto the top of 13% polyacrylamide gels (1.5 × 255 × 196 mm). Roti®-Mark PRESTAINED molecular weight marker (Roth, Karlsruhe, Germany) was used as a mass reference. After electrophoresis, the gels were incubated overnight in staining solution with colloidal Coomassie Brilliant Blue G-250. Quantitative analysis of individual groups of proteins was carried out in Image Master 2D Platinum software (GE Healthcare) using %Vol parameter (a ratio of the volume of a particular spot to the total volume of all spots present in the gel). The final result is an average of the spots %Vol obtained from three independent gels (technical repeats). In overall, about 200 samples were collected from 80 visible spots. Small spots were excised once, and thus each one contained a single sample. In turn, larger spots constituted for several samples due to multiple excision in different regions of the spot.

All samples were digested using Sequencing Grade Modified Trypsin (Promega, Madison, WI, USA). After digestion stage every sample was mixed in 1:1 ratio with the matrix. The matrix consisted of α-cyano-4-hydroxycinnamic acid diluted in 50% acetonitrile with 0.1% trifluoroacetic acid. The obtained peptide mixtures were analyzed on MALDI-ToF/ToF MS (Autoflex Speed, Bruker Daltonics, Billerica, MA, USA). The spectrometer was working in positive ions mode with the reflectron. The analysed ion masses ranged between 700 and 3500 Da. Calibration of the spectrometer was carried out every four samples, using standards in the range of analyzed peptides (Peptide Calibration Standards II, Bruker Daltonics). The obtained mass spectra were compared to those present in SwissProt database (The UniProt Consortium, www.uniprot.org) with the use of Mascot software. The search parameters included: mass tolerance: 0.25 Da, one incomplete cleavage allowed, alkylation of cysteine by carbamidomethylation (fixed modification), and oxidation of methionine (variable modification). Moreover, some peptides were selected for analysis in MS/MS mode. The peptides were sequenced by laser-induced dissociation (LID) using LIFT ion source. The search parameters for MS/MS data included: mass tolerance for MS mode: 0.25 Da, mass tolerance for MS/MS mode: 0.5 Da, one incomplete cleavage allowed, alkylation of cysteine by carbamidomethylation (fixed modification), and oxidation of methionine (variable modification).

Supplementary Materials: The following are available online, Annotated MS/MS spectra of the identified proteins.

Author Contributions: A.B. performed 2-D electrophoresis, protein identification, and wrote the manuscript; A.Ł. and K.K.H. performed bioinformatic analysis; J.B. performed 2-D electrophoresis and prepare samples for MS analysis; J.L. coordinated experiment; V.P. caught animals in their natural habitat; M.P. assisted with handling and fixating the animals, as well as collecting snake venoms.

Conflicts of Interest: The authors declare no conflict of interest.

References

1. Wüster, W.; Broadley, D.G. Get an eyeful of this: A new species of giant spitting cobra from eastern and north-eastern Africa (Squamata: Serpentes: Elapidae: *Naja*). *Zootaxa* **2007**, *1532*, 51–68, E-ISSN 1175-5334.
2. Warrell, D.A.; Greenwood, B.M.; Davidson, N.M.; Ormerod, L.D.; Prentice, C.R. Necrosis, haemorrhage and complement depletion following bites by the spitting cobra (*Naja nigricollis*). *Q. J. Med.* **1976**, *45*, 1–22. [CrossRef] [PubMed]
3. Tilbury, C.R. Observations on the bite of the Mozambique spitting cobra (*Naja mossambica mossambica*). *S. Afr. Med. J.* **1982**, *61*, 308–313, E-ISSN 2078-5135. [PubMed]
4. Warrell, D.A. Clinical toxicology of snakebite in Africa and the Middle East/Arabian Peninsula. In *Handbook of Clinical Toxicology of Animal Venoms and Poisons*; Meier, J., White, J., Eds.; CRC Press: Boca Raton, FL, USA, 1995; pp. 433–492. ISBN 9780849344893.
5. World Health Organization. *Guidelines for the Prevention and Clinical Management of Snakebite in Africa*; WHO/AFR/EDM/EDP/10.01; WHO, Regional Office for Africa: Brazzaville, Congo, 2010.

6. Warrell, D.A.; Ormerod, L.D. Snake venom ophthalmia and blindness caused by the spitting cobra (*Naja nigricollis*) in Nigeria. *Am. J. Trop. Med. Hyg.* **1976**, *25*, 525–529. [CrossRef] [PubMed]
7. Petras, D.; Sanz, L.; Segura, Á.; Herrera, M.; Villalta, M.; Solano, D.; Vargas, M.; León, G.; Warrel, D.A.; Theakston, R.D.; et al. Snake venomics of African spitting cobras: Toxin composition and assessment of congeneric cross-reactivity of the pan-African EchiTAb-Plus-ICP antivenom by antivenomics and neutralization approaches. *J. Proteome Res.* **2011**, *10*, 1266–1280. [CrossRef] [PubMed]
8. Rivel, M.; Solano, D.; Herrera, M.; Vargas, M.; Villalta, M.; Segura, Á.; Arias, A.S.; León, G.; Gutiérrez, J.M. Pathogenesis of dermonecrosis induced by venom of the spitting cobra, *Naja nigricollis*: An experimental study in mice. *Toxicon* **2016**, *119*, 171–179. [CrossRef] [PubMed]
9. Kini, R.M.; Doley, R. Structure, function and evolution of three-finger toxins: Mini proteins with multiple targets. *Toxicon* **2010**, *56*, 855–867. [CrossRef] [PubMed]
10. Konshina, A.G.; Boldyrev, I.A.; Utkin, Y.N.; Omel'kov, A.V.; Efremov, R.G. Snake cytotoxins bind to membranes via interactions with phosphatidylserine headgroups of lipids. *PLoS ONE* **2011**, *6*, e19064. [CrossRef] [PubMed]
11. Konshina, A.G.; Dubovskii, P.V.; Efremov, R.G. Structure and dynamics of cardiotoxins. *Curr. Protein Pept. Sci.* **2012**, *13*, 570–584. [CrossRef] [PubMed]
12. Dubovskii, P.V.; Konshina, A.G.; Efremov, R.G. Cobra cardiotoxins: Membrane interactions and pharmacological potential. *Curr. Med. Chem.* **2014**, *21*, 270–287. [CrossRef] [PubMed]
13. Gasanov, S.E.; Dagda, R.K.; Rael, E.D. Snake venom cytotoxins, phospholipase A_2s, and Zn^{2+}-dependent metalloproteinases: Mechanisms of action and pharmacological relevance. *J. Clin. Toxicol.* **2014**, *4*, 1000181. [CrossRef] [PubMed]
14. Feofanov, A.V.; Sharonov, G.V.; Dubinnyi, M.A.; Astapova, M.V.; Kudelina, I.A.; Dubovskii, P.V.; Rodionov, D.I.; Utkin, Y.N.; Arseniev, A.S. Comparative study of structure and activity of cytotoxins from venom of the cobras *Naja oxiana*, *Naja kaouthia*, and *Naja haje*. *Biochem. Mosc.* **2004**, *69*, 1148–1157. [CrossRef]
15. Feofanov, A.V.; Sharonov, G.V.; Astapova, M.V.; Rodionov, D.I.; Utkin, Y.N.; Arseniev, A.S. Cancer cell injury by cytotoxins from cobra venom is mediated through lysosomal damage. *Biochem. J.* **2005**, *390*, 11–18. [CrossRef] [PubMed]
16. Yang, S.H.; Chien, C.M.; Lu, M.C.; Lu, Y.J.; Wu, Z.Z.; Lin, S.R. Cardiotoxin III induces apoptosis in K562 cells through a mitochondrial-mediated pathway. *Clin. Exp. Pharmacol. Physiol.* **2005**, *32*, 515–520. [CrossRef] [PubMed]
17. Gomes, A.; Choudhury, S.R.; Saha, A.; Mishra, R.; Giri, B.; Biswas, A.K.; Debnath, A.; Gomes, A. A heat stable protein toxin (drCT-I) from the Indian Viper (*Daboia russelli russelli*) venom having antiproliferative, cytotoxic and apoptotic activities. *Toxicon* **2007**, *49*, 46–56. [CrossRef] [PubMed]
18. Chien, C.M.; Yang, S.H.; Chang, L.S.; Lin, S.R. Involvement of both endoplasmic reticulum-and mitochondria-dependent pathways in cardiotoxin III-induced apoptosis in HL-60 cells. *Clin. Exp. Pharmacol. Physiol.* **2008**, *35*, 1059–1064. [CrossRef] [PubMed]
19. Das, T.; Bhattacharya, S.; Biswas, A.; Gupta, S.D.; Gomes, A.; Gomes, A. Inhibition of leukemic U937 cell growth by induction of apoptosis, cell cycle arrest and suppression of VEGF, MMP-2 and MMP-9 activities by cytotoxin protein NN-32 purified from Indian spectacled cobra (*Naja naja*) venom. *Toxicon* **2013**, *65*, 1–4. [CrossRef] [PubMed]
20. Wu, M.; Ming, W.; Tang, Y.; Zhou, S.; Kong, T.; Dong, W. The anticancer effect of cytotoxin 1 from *Naja atra* Cantor venom is mediated by a lysosomal cell death pathway involving lysosomal membrane permeabilization and cathepsin B release. *Am. J. Chin. Med.* **2013**, *41*, 643–663. [CrossRef] [PubMed]
21. Barrington, P.L.; Yang, C.C.; Rosenberg, P. Cardiotoxic effects of *Naja nigricollis* venom phospholipase A_2 are not due to phospholipid hydrolytic products. *Life Sci.* **1984**, *35*, 987–995. [CrossRef]
22. Stefansson, S.; Kini, R.M.; Evans, H.J. The basic phospholipase A_2 from *Naja nigricollis* venom inhibits the prothrombinase complex by a novel nonenzymatic mechanism. *Biochemistry* **1990**, *29*, 7742–7746. [CrossRef] [PubMed]
23. Gowda, T.V.; Middlebrook, J.L. Effect of myonecrotic snake venom phospholipase A_2 toxins on cultured muscle cells. *Toxicon* **1993**, *31*, 1267–1278. [CrossRef]
24. Kini, R.M. Structure-function relationships and mechanism of anticoagulant phospholipase A_2 enzymes from snake venoms. *Toxicon* **2005**, *45*, 1147–1161. [CrossRef] [PubMed]

25. Montecucco, C.; Gutiérrez, J.M.; Lomonte, B. Cellular pathology induced by snake venom phospholipase A$_2$ myotoxins and neurotoxins: Common aspects of their mechanism of action. *Cell. Mol. Life Sci.* **2008**, *65*, 2897–2912. [CrossRef] [PubMed]

26. Lomonte, B.; Angulo, Y.; Sasa, M.; Gutiérrez, J.M. The phospholipase A$_2$ homologues of snake venoms: Biological activities and their possible adaptive roles. *Protein Pept. Lett.* **2009**, *16*, 860–876. [CrossRef] [PubMed]

27. Doley, R.; Zhou, X.; Kini, R.M. Snake Venom Phospholipase A$_2$ Enzymes. In *Handbook of Venoms and Toxins of Reptiles*; Mackessy, S.P., Ed.; CRC Press: Boca Raton, FL, USA, 2010; pp. 173–205. ISBN 9780849391651.

28. Rodrigues, R.S.; Izidoro, L.F.; de Oliveira, R.J., Jr.; Sampaio, S.V.; Soares, A.M.; Rodrigues, V.M. Snake venom phospholipases A$_2$: A new class of antitumor agents. *Protein Pept. Lett.* **2009**, *16*, 894–898. [CrossRef] [PubMed]

29. Zouari-Kessentini, R.; Luis, J.; Karray, A.; Kallech-Ziri, O.; Srairi-Abid, N.; Bazaa, A.; Loret, E.; Bezzine, S.; El Ayeb, M.; Marrakchi, N. Two purified and characterized phospholipases A$_2$ from *Cerastes cerastes* venom, that inhibit cancerous cell adhesion and migration. *Toxicon* **2009**, *53*, 444–453. [CrossRef] [PubMed]

30. Chen, K.C.; Liu, W.H.; Chang, L.S. Taiwan cobra phospholipase A$_2$-elicited JNK activation is responsible for autocrine fas-mediated cell death and modulating Bcl-2 and Bax protein expression in human leukemia K562 cells. *J. Cell. Biochem.* **2010**, *109*, 245–254. [CrossRef] [PubMed]

31. Khunsap, S.; Pakmanee, N.; Khow, O.; Chanhome, L.; Sitprija, V.; Suntravat, M.; Lucena, S.E.; Perez, J.C.; Sánchez, E.E. Purification of a phospholipase A$_2$ from *Daboia russelii siamensis* venom with anticancer effects. *J. Venom. Res.* **2011**, *2*, 42–51, E-ISSN 2044-0324. [PubMed]

32. Murakami, T.; Kamikado, N.; Fujimoto, R.; Hamaguchi, K.; Nakamura, H.; Chijiwa, T. A [Lys^{49}] phospholipase A$_2$ from *Protobothrops flavoviridis* venom induces caspase-independent apoptotic cell death accompanied by rapid plasma-membrane rupture in human leukemia cells. *Biosci. Biotechnol. Biochem.* **2011**, *75*, 864–870. [CrossRef] [PubMed]

33. Rodrigues, V.M.; Marcussi, S.; Cambraia, R.S.; de Araújo, A.L.; Malta-Neto, N.R.; Hamaguchi, A.; Ferro, E.A.; Homsi-Brandeburgo, M.I.; Giglio, J.R.; Soares, A.M. Bactericidal and neurotoxic activities of two myotoxic phospholipases A$_2$ from *Bothrops neuwiedi pauloensis* snake venom. *Toxicon* **2004**, *44*, 305–314. [CrossRef] [PubMed]

34. Santamaría, C.; Larios, S.; Angulo, Y.; Pizarro-Cerda, J.; Gorvel, J.P.; Moreno, E.; Lomonte, B. Antimicrobial activity of myotoxic phospholipases A$_2$ from crotalid snake venoms and synthetic peptide variants derived from their C-terminal region. *Toxicon* **2005**, *45*, 807–815. [CrossRef] [PubMed]

35. Xu, C.; Ma, D.; Yu, H.; Li, Z.; Liang, J.; Lin, G.; Zhang, Y.; Lai, R. A bactericidal homodimeric phospholipases A$_2$ from *Bungarus fasciatus* venom. *Peptides* **2007**, *28*, 969–973. [CrossRef] [PubMed]

36. Samy, R.P.; Stiles, B.G.; Gopalakrishnakone, P.; Chow, V.T. Antimicrobial proteins from snake venoms: Direct bacterial damage and activation of innate immunity against *Staphylococcus aureus* skin infection. *Curr. Med. Chem.* **2011**, *18*, 5104–5113. [CrossRef] [PubMed]

37. Fox, J.W.; Serrano, S.M. Structural considerations of the snake venom metalloproteinases, key members of the M12 reprolysin family of metalloproteinases. *Toxicon* **2005**, *45*, 969–985. [CrossRef] [PubMed]

38. Gutierrez, J.M.; Rucavado, A.; Escalante, T.; Díaz, C. Haemorrhage induced by snake venom metalloproteinases: Biochemical and biophysical mechanisms involved in microvessel damage. *Toxicon* **2005**, *45*, 997–1011. [CrossRef] [PubMed]

39. Gutierrez, J.M.; Rucavado, A.; Escalante, T. Snake venom metalloproteinases. Biological roles and participation in the pathophysiology of envenomation. In *Handbook of Venoms and Toxins of Reptiles*; Mackessy, S.P., Ed.; CRC Press: Boca Raton, FL, USA, 2010; pp. 115–138. ISBN 9780849391651.

40. Li, S.; Wang, J.; Zhang, X.; Ren, Y.; Wang, N.; Zhao, K.; Chen, X.; Zhao, C.; Li, X.; Shao, J.; et al. Proteomic characterization of two snake venoms: *Naja naja atra* and *Agkistrodon halys*. *Biochem. J.* **2004**, *384*, 119–127. [CrossRef] [PubMed]

41. Bocian, A.; Urbanik, M.; Hus, K.; Łyskowski, A.; Petrilla, V.; Andrejčáková, Z.; Petrillová, M.; Legáth, J. Proteomic Analyses of *Agkistrodon contortrix contortrix* Venom Using 2D Electrophoresis and MS Techniques. *Toxins* **2016**, *8*, 372. [CrossRef] [PubMed]

42. Nielsen, V.G.; Frank, N.; Matika, R.W. Carbon monoxide inhibits hemotoxic activity of Elapidae venoms: Potential role of heme. *BioMetals* **2018**, *31*, 51–59. [CrossRef] [PubMed]

43. Bocian, A.; Urbanik, M.; Hus, K.; Łyskowski, A.; Petrilla, V.; Andrejčáková, Z.; Petrillová, M.; Legáth, J. Proteome and peptidome of *Vipera berus berus* venom. *Molecules* **2016**, *21*, 1398. [CrossRef] [PubMed]

44. Tan, C.H.; Tan, K.Y.; Fung, S.Y.; Tan, N.H. Venom-gland transcriptome and venom proteome of the Malaysian king cobra (*Ophiophagus hannah*). *BMC Genom.* **2015**, *16*, 687. [CrossRef] [PubMed]

45. Trummal, K.; Samel, M.; Aaspõllu, A.; Tõnismägi, K.; Titma, T.; Subbi, J.; Siigur, J.; Siigur, E. 5′-Nucleotidase from *Vipera lebetina* venom. *Toxicon* **2015**, *93*, 155–163. [CrossRef] [PubMed]

46. Vogel, C.W.; Fritzinger, D.C. Cobra venom factor: Structure, function, and humanization for therapeutic complement depletion. *Toxicon* **2010**, *56*, 1198–1222. [CrossRef] [PubMed]

47. Terpinskaya, T.I.; Ulashchik, V.S.; Osipov, A.V.; Tsetlin, V.I.; Utkin, Y.N. Suppression of Ehrlich carcinoma growth by cobra venom factor. *Dokl. Biol. Sci.* **2016**, *470*, 240–243. [CrossRef] [PubMed]

48. Reichardt, L.F. Neurotrophin-regulated signalling pathways. *Philos. Trans. R. Soc. Lond. B. Biol. Sci.* **2006**, *361*, 1545–1564. [CrossRef] [PubMed]

49. Wijeyewickrema, L.C.; Gardiner, E.E.; Gladigau, E.L.; Berndt, M.C.; Andrews, R.K. Nerve growth factor inhibits metalloproteinase-disintegrins and blocks ectodomain shedding of platelet glycoprotein VI. *J. Biol. Chem.* **2010**, *285*, 11793–11799. [CrossRef] [PubMed]

50. Walsh, E.M.; Kim, R.; Del Valle, L.; Weaver, M.; Sheffield, J.; Lazarovici, P.; Marcinkiewicz, C. Importance of interaction between nerve growth factor and α9β1 integrin in glial tumor angiogenesis. *Neuro-Oncology* **2012**, *14*, 890–901. [CrossRef] [PubMed]

51. Osipov, A.V.; Terpinskaya, T.I.; Kryukova, E.V.; Ulaschik, V.S.; Paulovets, L.V.; Petrova, E.A.; Blagun, E.V.; Starkov, V.G.; Utkin, Y.N. Nerve growth factor from cobra venom inhibits the growth of Ehrlich tumor in mice. *Toxins* **2014**, *6*, 784–795. [CrossRef] [PubMed]

52. Dechant, G.; Barde, Y.A. The neurotrophin receptor p75[NTR]: Novel functions and implications for diseases of the nervous system. *Nat. Neurosci.* **2002**, *5*, 1131–1136. [CrossRef] [PubMed]

53. He, X.L.; Garcia, K.C. Structure of nerve growth factor complexed with the shared neurotrophin receptor p75. *Science* **2004**, *304*, 870–875. [CrossRef] [PubMed]

54. Salehi, A.; Delcroix, J.D.; Swaab, D.F. Alzheimer's disease and NGF signalling. *J. Neural Transm.* **2004**, *111*, 323–345. [CrossRef] [PubMed]

55. Tuszynski, M.H.; Thal, L.; Pay, M.; Salmon, D.P.; Bakay, R.; Patel, P.; Blesch, A.; Vahlsing, H.L.; Ho, G.; Tong, G.; et al. A phase 1 clinical trial of nerve growth factor gene therapy for Alzheimer disease. *Nat. Med.* **2005**, *11*, 551–555. [CrossRef] [PubMed]

Sample Availability: Samples are not available from the authors.

molecules

MDPI

Review

Application of Capillary Electrophoresis with Laser-Induced Fluorescence to Immunoassays and Enzyme Assays

Binh Thanh Nguyen [1,2] and Min-Jung Kang [1,2,*]

1   Molecular Recognition Research Center, Korea Institute of Science and Technology (KIST), Seoul 02792, Korea; 614007@kist.re.kr

2   Division of Bio-Medical Science and Technology (Biological Chemistry), Korea University of Science and Technology (UST), Daejeon 34113, Korea

*   Correspondence: mjkang1@kist.re.kr; Tel.: +82-2-958-5088

Academic Editors: Angela R. Piergiovanni and José Manuel Herrero-Martínez
Received: 24 April 2019; Accepted: 21 May 2019; Published: 22 May 2019

Abstract: Capillary electrophoresis using laser-induced fluorescence detection (CE-LIF) is one of the most sensitive separation tools among electrical separation methods. The use of CE-LIF in immunoassays and enzyme assays has gained a reputation in recent years for its high detection sensitivity, short analysis time, and accurate quantification. Immunoassays are bioassay platforms that rely on binding reactions between an antigen (analyte) and a specific antibody. Enzyme assays measure enzymatic activity through quantitative analysis of substrates and products by the reaction of enzymes in purified enzyme or cell systems. These two category analyses play an important role in the context of biopharmaceutical analysis, clinical therapy, drug discovery, and diagnosis analysis. This review discusses the expanding portfolio of immune and enzyme assays using CE-LIF and focuses on the advantages and disadvantages of these methods over the ten years of existing technology since 2008.

Keywords: CE-LIF; immunoassay; enzyme assay; chip-based CE-LIF assay

---

1. Introduction

Capillary electrophoresis (CE) has become an important tool in the era of separation since its first introduction by Jorgenson and Lukacs in 1981 [1]. The traditional technique, slab gel electrophoresis, initially demonstrated the application of electrophoresis, where charged molecules are separated under an applied electric field over the slab. Despite being straightforward and commonly used in numerous biological laboratories, slab gel electrophoresis is generally time-consuming, and has low efficiency and poor automation. Therefore, electrophoresis carried out in an open tubular glass capillary with an internal diameter of 75 µM was a momentous innovation concerning electrophoretic separation and the development of equipment and instrumentation later on [1,2].

Separation by CE can be conducted by several detectors. Presently, a vast number of detectors fall into one of two categories: bulk property or solute property detectors, where absorption detectors are specific to the latter and attribute to major commercial systems. Using UV or UV-VIS absorbance, the CE analysis deals with a universal range of bioanalytes, since most proteins and macromolecules, such as DNA or RNA, can absorb strongly at radiation in the UV or UV-VIS range [3–6]. Furthermore, the interface of CE to mass spectrometers or surface enhanced Raman spectroscopy (SERS) is rapidly being promoted to be an online tool to identify sample components [7–11].

Among the detection modes, laser-induced fluorescence (LIF) is one of the most sensitive techniques in terms of the determination and detection of a variety of biomolecules. In recent years,

CE has been established as an alternative method of conventional gel electrophoresis or in conjunction with high-performance liquid chromatography.

Bioanalysis based on immunoassays and enzymatic assays has gained a reputation in the field of biological studies and applications in pharmaceutical science, biomarker discovery, and clinical therapeutic and diagnostic targets [12–17]. A large number of studies have reported the practice of CE in the research of affinity binding between antibodies and antigens, or kinetic activities of different enzymes to further the understanding of many biological events and developing drug targets in the pharmaceutical industry [18–20]. A sophisticated analytical instrument for quantitative purposes such as CE-LIF has become emergent in the moving separation field of biomolecules [21–26]. Because CE-LIF performances are generally fast, automated, require a small number of samples, and are highly sensitive, they enable the simultaneous separation of various compounds at different sizes under minute records. Especially, CE-LIF can be merged into miniaturized systems that empowers it to be a high throughput, high-speed tool in the analysis of proteins and peptides [27]. Consequently, the use of CE-LIF in bioanalytical assays has drawn significant attention with the publication of numerous papers dealing with the analysis of biomolecules based on the two significant bio-reaction classes: (1) immune reaction and (2) enzyme reaction. The application of CE based on these two reactions has been extensively reviewed, summarized in [28–36].

In this article, we discuss the application of CE-LIF technique in the analysis of proteins and peptides, with an emphasis on immunoassays and enzyme assays in the last decade, from 2008 to early 2019. The details of the instrument conditions, method developments, and advances in the CE-LIF-based assay platforms in the biological studies are also reviewed.

2. CE-LIF Instrumentation Labeling Strategies for Peptides and Proteins Analysis

2.1. Instrumentation and Laser Sources

As suggested by its name, CE-LIF commonly uses lasers as its excitation source. To accomplish low limit of detection (LODs), it is crucial to maximize the signal and minimize stray light from the optical components and Raman scattering from the solvents. Instrumental LIF designs, such as orthogonal, epi-illumination, or sheath-flow cuvette, have been continually developed in line with CE systems to achieve high sensitivity.

LIF sources for peptide and protein analysis strongly link to the use of fluorescent dyes. These fluorophores absorb light energy in the range of 350–650 nm wavelength. In this range, gas lasers are the most common sources: He–Ne laser (543.5, 593.9, 632.8 nm), Ar laser (454.6, 488, 514.4 nm), Kr laser (416, 530.9, 568.2, 647.1 nm) [37]. Diode lasers provide a stable, thermoelectrically cooled, efficient performance and offer a higher frequency [38,39]. While gas lasers require a kilo-volts power, diode lasers typically run on small voltage supplies. Semiconductor diode lasers emit in the near-infrared (NIR) range, which is an attractive setting which offers a lifetime of over 10,000 h and a volatile range of wavelength [40–42]. Another emerging source is light emitting diodes (LEDs), which generate monochromatic light and are a source of great potential for fluoresce measurements in the miniaturized system, and have been used successfully for the detection of various biomolecules [43–47]. LED stands for light emitting diode, and they are a semiconductor device that emit visible light under the application of an electric current. The output from an LED can range from red (~700 nm wavelength) to blue-violet (~400 nm wavelength), or even the infrared (IR) region (~830 nm or longer). Covering a broad spectrum of emitting light, LEDs have become a reliable replacement for LIF, resulting in the formation of LEDIF detection (Light Emitting Diode Induced Fluorescence). CE-LEDIF has been used for the separation of various peptides and proteins [48–50]. LEDs are promising as they are less expensive, have lower energy consumption, are more stable, and have a longer lifetime compared to lasers.

LIF detectors can be integrated into an existing commercial CE system, or are available as a standalone, which can be fused with a range of commercial CE or HPLC. The principle of an integrated detector system is using a ball lens to level the laser beam onto the capillary window and an ellipsoid mirror glued on the capillary to collect the emitted fluorescence. The system is secured inside a CE cartridge. A standalone LIF detector typically consists of a laser module, a photomultiplier tube, and the optics with filters, and a dichroic mirror used to reflect the laser beam [51,52]. The flexibility of LIF detection settings allows a dynamic change during operations. For example, Dada et al. reported that LIF detectors, in conjunction with two fiber optic beam splitters and two avalanche photodiodes, gave satisfactory results in the determination of biological analytes in a wide dynamic range [53].

2.2. Labeling

CE analysis of proteins and peptides typically utilizes the absorbance detection, which takes advantage of the absorbance of peptide bonds in the ultraviolet (UV) region. For example, spectrophotometry of peptide bonds primarily responds at 200 nm wavelength, and amino acids with aromatic rings are responsible for the absorbance peak at 280 nm wavelength. LIF detection has advanced to be one of the most sensitive CE methods, as it is reported to detect the samples at attomole or zeptomole levels. The first and foremost prerequisite for an analyte to be detected by LIF is the need to be natively fluorescent or chemically derivatized with a designed fluorophore, which is excited at an appropriate wavelength responding to a laser source. Because many peptides and proteins do not exhibit sufficient native fluorescence, due to the dearth of intrinsic fluorescent amino acids, such as tyrosine, tryptophan, and/or phenylalanine, the use of derivatized peptides and proteins has become a common practice of LIF. However, the derivatization procedure contains some drawbacks. For example, at low concentrations of analytes, the reaction yield is low, resulting in poorly labeled analytes and high concentrations of fluorescent backgrounds. In addition, it is challenging to achieve accuracy and reproducibility for the labeling method, thus making the derivatization procedure undesirable for certain types of the analyte. Furthermore, another bottleneck with derivatization of complex analytes is the formation of products attached with multiple fluorophores, which produces a multi-peak chromatography, leading to a perplexing quantification. To circumvent these difficulties, the development of derivatization protocols or studies of new fluorescence dyes has been continuously ongoing. Derivatization of peptides and proteins can be performed pre-, on-, or post-column. Each of these methods has advantages and disadvantages depending on the nature of the fluorophore.

Among covalently and fluorescently labeled proteins and peptides, in most case, the dyes can react with primary and secondary amines of amino acids or with thiol groups of cysteine residues. A fluorogenic dye is understood as a precursor of a labeled protein or peptide of interest, but is poorly fluorescent in native form. However, it becomes strongly fluorescent in tagged proteins or peptides after undergoing a chemical or enzymatic reaction. Typical fluorogenic dyes, such as naphthalene-2,3-dicarboxylaldehyde (NDA), 5-furoylquinoline-3-carboxyaldehyde (FQ), and (4-carboxylbenzoyl) quinoline-2-carboxaldehyde (CBQCA), are widely used in LIF detection of the visible region (280–400 nm). LODs obtained are in the range μM–pM [54–58]. The chemical structures and labeling reactions of these dyes are shown in Figure 1. Other covalent labels include 4-Chloro-7-nitro-2,1,3-benzoxadiazole (NBD-Cl), 6-aminoquinolyl-*N*-hydroxysuccinimidyl carbamate (AQC), or near-infrared (NIR) dyes, for example NN382 (LICOR, Inc.).

Besides the utility of fluorogenic dyes, many studies describe fluorescent dyes, such as FITC and rhodamine-based dyes, for LIF detection of longer wavelengths (400–600 nm). Labeling with FITC has become attractive, since it is highly fluorescent [59,60]. However, it is less reactive and efficient at low concentrations of amines with LOD obtained in the range of μM. Nevertheless, rhodamine-based dyes are more reactive compared to FITC, especially if they are activated by the succimidyl ester group, resulting in an efficient reaction with unprotonated amines to form a stable amide bond, thus enabling the LODs in the pM range [61]. For instance, Korchane et al. presented a pre-capillary derivatization strategy of two synthetic transthyretin peptides for the pathology diagnosis, using two fluorogenic dyes (NDA and FQ) and one rhodamine-based tag 5-Carboxytetramethylrhodamine, Succinimidyl Ester (TAMRA-SE). They successfully separated the wild type and mutated type under optimal conditions, with TAMRA-SE labeled derivatives giving the highest resolution, whereas NDA displayed the best detection sensitivity (LOD of 2.5 µM) [62].

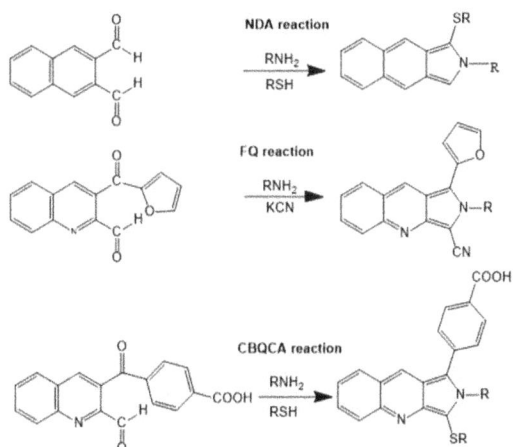

Figure 1. Chemical structures and labeling reactions of naphthalene-2,3-dicarboxylaldehyde (NDA), 5-furoylquinoline-3-carboxyaldehyde (FQ), and (4-carboxylbenzoyl) quinoline-2-carboxaldehyde (CBQA).

Parallel to the use of covalent dyes, noncovalent labels are presented as viable alternatives to reduce the sample handing steps. For instance, idocyanine green greatly fluoresences once it is noncovalently bound to protein, thus allowing the CE-LIF detection at 780 nm [63]. Indigocarmine blue is similar to idocyanine green—it absorbs and emits at 436 nm and 528 nm, respectively. Other noncovalent dyes include Nano Orange, Sypro red, Sypro orange, and Sypro tangerine, which could be detected efficiently under a 488 nm laser. CE-LIF analysis of biopolymers has benefited from using these noncovalent labels, regardless of the slow kinetics of the reactions. While these noncovalent dyes seem to be an interesting approach, the outputs are not conclusive for a broad range of proteins and peptides.

When selecting fluorescent reagents in CE-LIF practice, one needs to consider the following: high quantum yields with low quenching, rapid reaction rates, protein conjugate photostability, and derivatization homogeneity. For example, FQ reacts with ε-amines of lysine to form a stable fluorescent indole derivative, thus generating FQ–protein ligands with high quantum fluorescence efficiencies. Derivatization is accelerated at high temperatures and a moderate pH, and the reaction is simply quenched by dilution with SDS.

Bioassays using native fluorescent proteins and peptides will be excluded from this review, as we focus on the discussion of immunoassays and enzyme assays, as they both rely on the use of fluorescent antigens or antibodies and substrates.

3. CE-LIF-Based Immunoassays

Immunoassays using CE techniques—called capillary electrophoretic immunoassays (CEIA) or immunocapillary electrophoresis (ICE)—incorporate an immunological reaction in CE. Unlike other immunoassay methods, CEIA allows the direct visualization of the immunocomplex product formed between an antigen and an appropriate antibody, hence simplifying the interpretation of the results. Compared to traditional methods, such as ELISA, CEIAs are appealing to the study of bioanalytes due to their ease of automation and feasibility of trace amount detection of samples. The concept of performing an immunoreaction on a capillary was first introduced by Nielsen in 1991. Two years later, Schultz and Kennedy demonstrated CE-immunoassays on both competitive and noncompetitive formats to measure insulin using FITC labeled reagents. The combination of the specific immune reactivity with high separation efficiency of CE has rendered the use of this technique to numerous bioanalytes of interest.

Practically, CE immunoassays can be adapted to both homogenous and heterogenous systems. In homogenous assays, analyte, antibodies, and other reacting species are all present in the liquid phase. Two formats that are considered for the combination of CE with immunoreactions in the homogenous system are non-competitive binding and competitive binding, which will be discussed in detail below.

CEIA can be performed on various detection modes, from UV, LIF, and mass spectrometry. CEIA-LIF has become a preferred tool for the analysis of many research groups, due to the high sensitivity and selectivity of the method. The following sections describe the two formats used in CEIA-LIF.

3.1. Non-Competitive Binding Format Assay

3.1.1. Principle

Non-competitive CEIA or affinity probe capillary electrophoresis (APCE) involves the quantification of the immunocomplex, which is directly proportional to the number of labeled analytes. Either Ag or Ab needs to be fluorescently labeled. There are two different options for tagged reactants, where Ag* and Ab* indicate labeled reagents:

$$Ab + Ag* \text{ (excess)} \leftrightarrow Ab\text{-}Ag* + Ag* \text{ (excess),}$$

$$Ab* \text{ (excess)} + Ag \leftrightarrow Ab\text{-}Ag* + Ab* \text{ (excess).}$$

Note that an excess amount of the labeled reagent is added to ensure the complete binding of the analyte present in the reaction mixture. The LIF detector reveals the peak profiles of both the immunocomplex product and excessive labeled Ag or Ab based on their relative differences in size and charges. It is possible to quantify the amount of Ab or Ag in the mixture according to the amount of immunocomplex formed or/and the decreased amount of free labeled agents. Nevertheless, this direct format is limited in use because the binding between small molecules, as with an antigen, and large molecules, as with an antibody, do not significantly vary the electrophoretic mobility of the labeled molecular recognition elements, hence impeding the separation of labeled reagents from the immunocomplex.

3.1.2. Application

The non-competitive CEIA format coupled with LIF detection has been widely used as a part of APCE techniques in the separation of a large variety of biomolecules. The major advantage of non-competitive assays is the commercial availability of labeled antibodies. Liu et al. developed a non-competitive CEIA-LIF to detect alpha-fetoprotein (AFP) in the early diagnosis of primary hepatoma [64], using poly (guanidinium ionic liquid) monolithic material. In this assay, AFP was incubated with an excess amount of fluorescently labeled antibody to form an immunocomplex, and thereafter separated accordingly. Under optimized conditions, their assay performed an LOD of 0.05 μg/L AFP. Compared to other immunoassays to detect AFP, their method exhibited higher sensitivity and larger linear dynamic range, specifically, no purification process, thus shortening the analysis time. By combining a CE-LIF immunoassay with fluorescence polarization, Wang et al. described a method for rapid and sensitive detection of genomic DNA methylation without tedious processes such as the bisulfate conversion, enzyme digestion, or PCR amplification. In this assay, the immunocomplex of methylated DNA was recognized by the fluorescently labeled secondary antibody and separated from unbound antibody by CE-LIF. The analytical performance accomplished an LOD of DNA methylation at 0.3 nM, proving the CE feasibility in the separation of a great diversity of compounds [65]. Besides this, it demonstrated the flexibility of labeling reagents, not only limited to primary antibodies but also applicable to secondary antibodies.

As mentioned previously, the direct format is challenging to achieve because of unrecognized electrophoretic mobility change and difficulty in homogenous labeling. As a result, the use of labeled antibodies as detected traces has been restricted. Instead, several research groups are paying attention to the application of aptamers for use as a binding ligand in APCE separation. Aptamers are single-stranded nucleic acids with advantageous features over antibodies, such as high stability, ease of synthesis, and high binding capacity. In CE-LIF, the aptamer is usually labeled with a fluorescent dye. Owing to their smaller molecular weight than proteins (5–15 kDa), the binding of aptamers on larger size proteins can significantly change the ratio of charge to mass of the labeled aptamer. Consequently, the electrophoretic mobility change is significantly improved, enabling the CE separation of immunocomplex and free aptamers much more easily. Hao et al. described a non-competitive CE-LIF-based method to study human thrombin, an essential protein related to the blood coagulation pathway, and successfully detected thrombin in human serum at 0.2 nM using dye-labeled nuclease resistance aptamer Toggle-25 (Figure 2) [66]. Prior to their work, Song et al., reported an affinity probe capillary electrophoresis/laser-induced fluorescence polarization (APCE/LIFP) and achieved the detection of thrombin at low LOD at sub nanomoles/liter [67]. Yi et al. demonstrated a non-competitive CE-LIF assay to detect picomolar concentrations of glucagon and amyline, using labeled mirror-image aptamers called Spiegelmers [68]. The LOD obtained for glucagon was 6 pM and for amylin was 40 pM. These LOD values were lower compared to those obtained from the competitive immunoassay format using rival Ab as affinity ligands in their previous works. The reproducibility of the detection methods for glucagon and amylin (Relative Standard Deviation for peak height, RSD) were <5.8% and <5.3%, respectively.

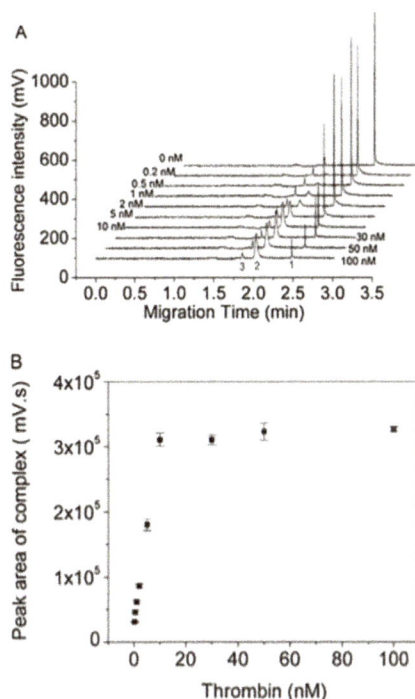

Figure 2. Capillary electrophoresis using laser-induced fluorescence detection (CE-LIF) detection of thrombin using tetramethylrhodamine TMR labeled Toggle-25. (**A**) Electropherograms of Toggle-25-TMR in the presence of varying concentrations of thrombin. (**B**) The relationship between the peak area of complex and the concentrations of thrombin. Reproduced under permission of [66].

By using aptamer CE-LIF application, other non-competitive assays have successfully detected several proteins and peptides, including platelet-derived growth factor (PDGF) [69], human immunodeficiency virus reverse transcriptase [69], human immunoglobulin E [69], and recombinant human erythropoietin-α [70]. CE-LIF aptamer separation also enables the simultaneous analysis of multiple proteins in a single run, ranging from various proteins [69] to small analytes [71]. In most aptamer-based affinity CE-LIF assays, DNA aptamers are usually favored, whereas the use of RNA aptamer is still modest. This is due to the poor stability of RNA aptamers compared to DNA aptamers. Especially in the biological matrix, RNA aptamers tend to be quickly degraded. When selecting a non-competitive format immunoassay, one needs to consider several factors to maximize the performance. For instance, only tagged species can produce fluorescence, and these signals should not be interfered with by analytes or sample matrix. The sample matrix or background electrolyte effect should be studied carefully, because they can enhance or reduce the signals which confound the legitimacy of the method. The compilation of the non-competitive format assay is presented in Table 1.

Table 1. Recent CE-LIF immunoassays.

Analyte	Format	Labeled	LOD	Ref.
CCP peptides	Competitive	FITC	4 ng/mL	[60]
Alpha-fetoprotein	Non-competitive	FITC	0.05 µg/mL	[64]
DNA fragments	Non-competitive	Alexa Fluor 546	0.3 nM	[66]
Thrombin	Non-competitive	TMT aptamer	0.2 nM	[66]
Thrombin	Non-competitive	TMT aptamer	0.2 nM	[67]
Glucagon	Non-competitive	6-FAM aptamer	6. 0 pM	[68]
amyline	Non-competitive	6-FAM aptamer	40 pM	[68]
IgE	Non-competitive	5-FAM	250 pM	[69]
human immunodeficiency virus reverse transcriptase	Non-competitive	5-FAM	100 pM	[69]
PDGF-BB	Non-competitive	5-FAM	50 pM	[69]
Recombinant human erythropoietin-α	Non-competitive	FITC	0.2 nM	[70]
human serum albumin	Competitive	FITC	1.34×10^{-7} M^{-1}	[72]
norfloxacine	Competitive	FITC	0.005 µg/L	[73]
carbaryl	Competitive	FITC	0.05 ng/mL	[74]
chloramphenicol	Competitive	FITC	0.0016 µg/L	[75]
receptor beta- pdgf	Competitive	6'-FAM	3 nM	[76]
receptor alpha- pdgf	Competitive	6'-FAM	0.5 nM	[76]
testosterone	Competitive	FITC	1.1 ng/mL	[77]
chloramphenicol	Competitive	FITC	7.6×10^{-9} g mL^{-1}	[78]
glucagon	Competitive	FITC	5 mM	[79]

3.2. Competitive Binding Format Assay

3.2.1. Principle

In a traditional competitive immunoassay, the amount of one reactant is limited. Similarly, in a CEIA-LIF-based assay, the fluorescent reagent (Ag* or Ab*) competes with non-fluorescent analog (Ag or Ab) to bind with a limited amount of corresponding immunological reactants. The reactions can be formulated as follows:

$$Ag + Ag^* + Ab \text{ (limited)} \leftrightarrow Ab\text{-}Ag + Ab\text{-}Ag^* + Ag + Ag^*,$$

$$Ab + Ab^* + Ag \text{ (limited)} \leftrightarrow Ab\text{-}Ag + Ab^*\text{-}Ag + Ab^* + Ab.$$

The fluorescently labeled reactant competes with free isotypes in the mixture to form a fluorescent immunocomplex. CE-LIF profiles display two distinct peaks, corresponding to the free labeled reactant and the immunocomplex. The concentration of the antigen (or antibody) is directly proportional to that of the labeled antigen (or antibody), but inversely proportional to that of the immunocomplex.

3.2.2. Application

Since being developed, the competitive binding assay has gained a growing reputation and become the most popular format for CE-LIF-based immunoassays. Giovannoli et al. assessed the variables affecting the performances of a competitive CE-LIF assay for the detection of human serum albumin. They showed that the development of the CEIA-based assay was similar to that applied in conventional microplate immunoassays, regarding the interface of peculiarities [72]. The same group also reported another CE-LIF competitive immunoassay to investigate Cry1bAb endotoxin from Bacillus thuringensis with an LOD of 0.5 nM, and allowed the satisfactory recovery (62–98%) of the protein in real samples [80]. In 2018, we successfully developed a quantitative CE-LIF-based assay to detect the antibodies against cyclic citrullinated peptides (CCP) for the diagnosis of rheumatoid arthritis. Our method enabled the quantification of the concentration of anti-CCP antibodies in patient sera, ranging from 0.1–0.4 µg/mL (Figure 3). For the detection of anti-CCP antibodies, our assay achieved a reproducibility within 5% and accuracy ranging between 89% and 103%. Compared to a semi-quantification method, such as ELISA, our CE-LIF-based method outperformed ELISA in terms of specificity and sensitivity [60]. In addition, the analysis was fast, reproducible, and no clean up step was required for intricate sample matrices, making it highly adaptable to other disease diagnoses using biological fluids. In 2015, Liu et al. discussed a simple competitive CE-LIF assay to determine the concentration of norfloxacin in food samples. Norfloxacin is a compound that is widely used in the treatment of gonococcal urethritis, respiratory and skin infections, but is consequently retained in animal-derived foods and leads to public health threats. Compared to the traditional methods such as ELISA and HPLC-MS/MS, their CEIA-LIF displayed high sensitivity, reduced laborious washing steps, and established an efficient quantitative tool for the selective determination of chemical residues in biological matrices. The proposed method obtained an LOD of 0.005 µg/L of norfloxacin, and the RSDs for migration time and peak area of the immunocomplex were 0.17% (intraday) and 3.46% (interday), respectively [73]. Zhang et al. performed a CE-LIF competitive format to effectively detected carbaryl in rice samples, with 14 times greater sensitivity than that of ELISA, which used the same immune-reagents [74]. Several groups also reported the application of CEIA-LIF in the detection of alpha-fetoprotein (AFP) and thyroxine (T4) in human sera, and chloramphenicol in animal-derived foods [75].

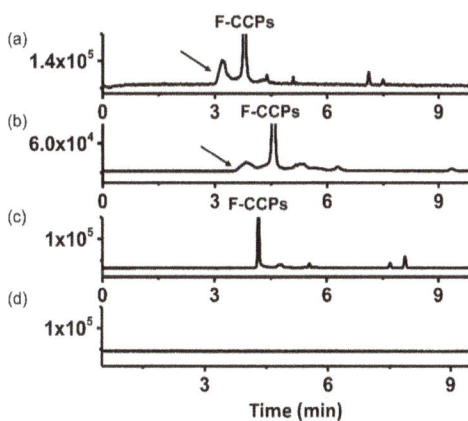

Figure 3. Electropherograms of immunocomplexes of fluorescent cyclic citrullinated peptides (F-CCP) with (**a**) anti-cyclic citrullinated peptides (CCP) antibodies in PBS, (**b**) anti-CCP antibodies in Fetal Bovine Serum, and (**c**) human IgG. (**d**) Electropherogram of patient samples without F-CCP treatment. Arrows indicate the peak from the immunocomplex of F-CCP and anti-CCP antibodies. Reproduced with the permission of [60].

Like the non-competitive format, CEIA-LIF competitive assays benefit enormously from the development in the application of various bioanalyses. Zhang et al. introduced a method to determine receptors of PDGF, where the receptor β competed with a fluorescent aptamer in binding to PDGF-BB.

The aptamer bound strongly with the β receptor but not the α receptor of PDGF, resulting in a difference in the electrophoretic mobility of PDGF isomers, which allowed the separation of PDGF's isomers in a single analysis. This is also a significant advantage of CE over MS which cannot distinguish compounds with the same molecular weight. Zhang's work successfully demonstrated the simultaneous detection of PDGF isomers and their receptors in a single run [76]. Aptamers binding to analytes often leads to signal changes, a mechanism called structure switching. Using this concept, Zhu et al. described a novel APCE strategy dedicated to small molecule detection. They established a pure, high throughput and versatile APCE-LIF in terms of adaptability, generalizability, and capability for multiple detections of small size analytes in a single capillary [81]. Other applications of competitive binding CE-LIF aptamer-based immunoassays can be found in Table 1.

Competitive binding assays have become the most critical platform in CE-LIF-based immunoassays, due to the easier separation of the bound and non-bound labeled reagents. Furthermore, only an analog of the analytes needs to be labeled, which prevents the production of multiple homogeneous preparations of labeled antibodies that are often encountered in the non-competitive format. However, competitive assays tend to have a higher limit of detection and a smaller dynamic range compared to that of non-competitive assays, and more difficulty in distinguishing cross reactivity between species than when a non-competitive format is employed.

3.3. Microchip-Based CEIA-LIF

Electrophoresis performed in microchips was first introduced by Manz and Harrison in 1992 [82]. Compared to traditional CE, the integration of microfluidics into the system to manipulate, automate, and analyze the minimum volume of analytes has opened up a new era for high-throughput analysis, a key to industrialized drug discovery. Today, many pharmaceutical companies are screening up to 300,000 or more compounds per screen to produce 100–300 hits. The concept of immobilizing immuno reagents for measuring the bioactivities of drugs adds a dimension to the existing drug discovery paradigm. Numerous reviews have discussed the application of microfluidic devices in the CE system with regards to the time period [83–87]. The previous sections discuss a general homogenous format, where antibodies and antigens are present in the solution phase. The chip-based method represents a heterogeneous format in which either the analyte, antibody, or analog of the binding agents are immobilized onto a solid support [88]. Similar to the homogenous system, immunoassays carried out by CE use both noncompetitive and competitive formats [89,90]. The combination with LIF high selectivity and sensitivity features has drawn attention to microchip-based CE-LIF for use in biological and clinical studies. Phillips and Wellner introduced an immunoaffinity chip-based CE-LIF method to measure the concentration of a brain-derived neurotrophic factor in human skin biopsies [91]. The antibodies were chemically immobilized on a replaceable immunoaffinity disk. Homogenates obtained from micro-dissected human skin samples were subject to the immunoaffinity inserts, in which the analyte of interest was captured, followed by fluorescent labeling with a red-emitting laser dye, before being detected by LIF. Compared to conventional immunoassays, this chip-based CEIA-LIF demonstrated a good correlation. In addition, this system has the potential to be modified into a portable unit for clinical or biomedical screening. The same group later designed another microchip-based CE-LIF device to study chemokines in samples of neuro inflammatory premature infants. Using similar strategies as in the previously described method, the system took only two minutes to successfully isolate six analytes in a single run. The method compared well to a commercial ELISA. In addition, the CE chip was more reliable, and required significantly fewer samples, a crucial criterion when studying newborns [92]. Very recently, this group published a chip-based CE-LIF immunoassay to study inflammatory mediators in newborn dried blood spot samples [93].

The application of chip-based CE immunoassays to multi-analysis was demonstrated by the Shi et al. study on multiple tumor markers [94]. The detection of carcinoma antigen 125 (CA125) and carbohydrate antigen 15-3 (CA15-3) was based on an offline noncompetitive immunoreaction of CA125 and CA15-3 with FITC-labeled monoclonal antibodies. Subsequently, the microfluidic multiplexed channel coupled with the LIF detector allowed the separation of CA125 and CA15-3 contents in the sera of cancer patients, as well as healthy cohorts. In contrast to various traditional immunoassays determining CA125 and CA15-3, such as immunoradiometric assays or immunofluorometric assays, the chip device enabled a rapid, small reagent consumption, and simple operation performance.

Chip-based CEIA-LIF relies on the sources in which the reagents are immobilized. For the methods based on immobilized antibodies, the conditions should be considered cautiously in order to release the captured analytes, which sometimes remain challenging because some antibodies with strong affinity binding are utilized for such work. However, because of its strong binding capacity, this approach is useful to isolate and concentrate trace substances before analysis by CE. It is necessary to ensure there is no irreversible damage to the immobilized antibodies on the chip. For the methods based on immobilized analogs of the analyte, to create the immobilized support, the analyte may need to be derivatized with appropriate functional groups, which help the immobilization within the CE wall. Moreover, this procedure should not interfere with the interaction between antibodies and antigens. For a small-size immobilized tracer, a spacer arm must be included to link the analog and the device, to enable accessibility to antibodies for binding.

4. CE-LIF-Based Enzymatic Assays

Since first being introduced by Banke et al. [95] to detect the enzyme activity of alkaline protease, CE coupled with enzymatic reaction has been widely applied to study and characterize enzyme-catalyzed analysis. Enzymes regulate nearly all physiological chemical reactions in living organisms and catalyze all aspects of cell metabolism. The extensive studies of enzymes as current drug targets have encouraged researchers to seek effective methods to characterize enzyme activities and understand their roles in human diseases. Such activities include enzyme kinetics, enzyme substrate identification, enzyme inhibitor screening, and enzyme-mediated metabolic pathways, and these aspects of enzyme-related analysis have been largely studied by CE methods, for example, those summarized in [96–99]. A typical CE-based enzyme assay starts with small amounts of the substrate, enzyme, and/or inhibitors. When the enzymatic reaction is completed, the CE can directly detect and separate the components of interest without the requirement of additional coupling factors, such as co-enzyme, which reduce the sample complexity and the likelihood of fall positives, especially when it comes to the studies of enzyme inhibition. In general, CE-based enzyme assays can be grouped into two main categories: (1) pre-capillary (offline) assays, in which the reaction is performed outside the capillary before being injected into the system, and (2) in-capillary (online) assays, in which the enzymatic reaction takes place inside the capillary, where injection, mixing, reaction, separation, and detection are integrated into a single column. Details of these two categories are discussed in this section.

4.1. Off-Column (Pre-Column) Enzymatic Assays

In off-column assays, or so-called pre-column enzymatic assays, the reaction is carried out in a separate system. Enzymes, substrates, and/or inhibitors are mixed and incubated, and the sample is subjected to CE analysis, where the substrates' and/or inhibitors' activities are determined. The capillary only functions as a separating channel. Because this format is easy to monitor—due to the ability of enzymes to catalyze under straightforward conditions—offline methods have been applied to enzyme inhibitor screening and drug metabolism studies for years. However, there are some drawbacks to the offline mode. Firstly, CE consumes only small volumes of the sample. However, since it is performed in a vial, the reaction requires much larger volumes of both enzymes and substrates to undergo the reaction and multiple steps may be needed to operate, such as adding the quenching reagents to terminate the enzymatic reaction, or changing the reaction conditions before the analysis by the CE system. This leads to the waste of reagents, especially for expensive enzymes, and the addition of strong acids during the quenching step may contribute to the peak distortion later. Moreover, interference between the incubation buffer and separation background electrolyte may result in EOF variability, peak broadening, and eventually low sensitivity and reproducibility of the method. Therefore, the LIF mode has offered to enhance the sensitivity for the detection based on pre-capillary methods.

4.2. Application

In 2016, we published a pre-capillary CE-LIF assay for the inhibitor screening of protein kinase C [59]. Our method used FITC labeled ERK peptides F-ERK and P-F-ERK (Extracellular signal Regulated Kinases) to act as Protein Kinase C δ substrates and the enzymatic reaction was conducted in a cellular system (Figure 4). Prior to the enzyme assay, our method accomplished LODs of 4 and 12 ng/mL for F-ERK and P-F-ERK, respectively. The reproducibility for the two peptides was within 5% and accuracy ranged between 86% and 109%. We then successfully calculated the IC50 values of four inhibitors of PKCδ, including staurosporin, bisindolylmaleimide II, gö6983, and rottlerin. Compared to a commercial PKC ELISA kit, our assay provided the exact quantification and was more adaptable to differing enzyme isoforms. Lee et al. described a pre-capillary CE-LIF method to determine the kinase activity of sphingosine for application in preclinical and clinical trials [100]. Their assay allowed the determination of the in vitro activity of both kinase and phosphates, using purified enzymes. While the traditional platforms to study sphingosine activity, like radiometric assays, contained many drawbacks, such as limited sensitivity, semi-quantitative results, poor resolution, and being time-consuming, the CE-LIF-based assay offered a quantitative tool, high sensitivity, and a robust and straightforward method, which was amenable for the study of enzymes of interest in both cell biology and clinical medicine.

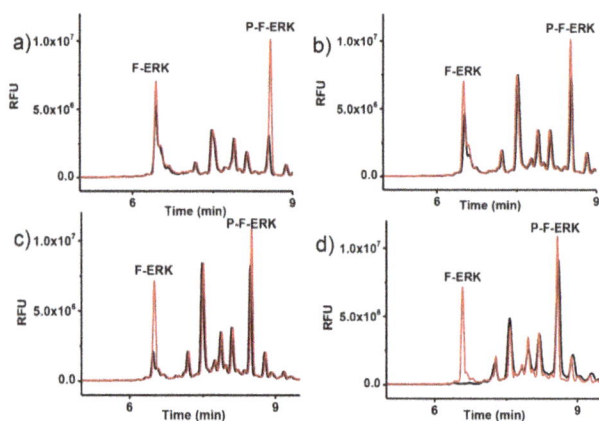

Figure 4. Time course study of enzymatic reaction. Extracted media from gastric cancer MKN-1 cells treated with 2 μg/mL F-ERK, 0.01 μg/mL PKCδ, 50 mM Adenosine triphosphate ATP, and incubated at different time intervals. (**a**) 3 h incubation; (**b**) 6 h incubation; (**c**) 12 h incubation; (**d**) 24 h incubation. Reproduced with the permission of [59].

DNA demethylation is essential for organism survival and the enzymes that catalyze the reaction have been studied in several bodies of research. However, most approaches investigated in vitro only indirectly via the detection of coproducts. Karkhanina et al. described a method to directly measure the formation of the demethylated DNA product, using a 15 nt long one-base methylated substrate, with the separation only taking 10 min [101]. While separating DNA typically uses the gold standard technique slab gel electrophoresis, using such traditional methods failed to perform, due to the lack of difference in the lengths and conformation of the product and substrate. However, CE successfully achieved well-resolved peaks of both the product and the substrates.

Histone deacetylation plays a vital role in gene expression, and aberrant transcription due to mutated genes that encode histone deacetylaes (HDAC) is a hallmark in the onset progression of cancer [102,103]. Hence, inhibitors of HDACs have emerged as a new class of drugs for the treatment of cancers, because of their effects on tumor suppression, cell growth, and cell survival. Recently, Zhang et al. established a rapid and cost-effective CE-LIF based method, employing a 5-carboxyflyrescein labeled peptide with an acetylated lysine residue as the substrate of HDAC1 to screen the HDAC inhibitors from 38 purified natural products [104]. Their calculated IC50 value for a well-known HDAC inhibitor—suberoylanilide hydroxamic acid SAHA—was consistent with that of the literature. From the screened products, luteolin was identified as an HDAC inhibitor. The method provided herein strengthened the ability of CE-LIF as a universal approach for the screening of many other kinds of enzyme inhibitors.

Picard et al. introduced a platform for the profiling of multiple proteolytic activities, using a fluorescent-labeled substrate assaying in the absence and presence of protease inhibitors [105] (Figure 5). Using a commercially available 96-channel capillary DNA sequence, coupled with CE-LIF, they successfully demonstrated the monitoring, classification, and inhibition of multiple proteolytic activities acting on the model peptide—an amino acid sequence of mouse granulocyte chemotactic protein-2. Numerous biological protease mixtures, including proteases from tumor cells, neutrophil granulocytes, and plasma, were studied. Although their method could be a high throughput starting point to investigate relevant proteases in cells and in vivo, a critical drawback is that the peptide model studied may not reflect the exact conformation within the context of the protein, therefore implying a substantial impact of the capability of proteases to cleave peptide compared to protein-based substrates.

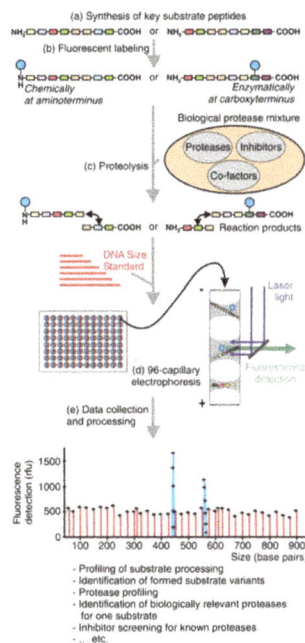

Figure 5. Microfluidic devices used for parallel electrophoretic enzyme assays. (**a**) Design of microfluidic network containing 16 parallel separation channels, (**b**) design of a 36-channel network, (**c**) photograph of a finished 36-channel chip, (**d**) bright-field image of the detection area on the 36-channel chip, (**e**) repetitive units of the microfluidic network. The Electrokinetic injection procedure is described in experimental section. Reproduced with the permission of [105].

In 2018, Fayad et al. successfully demonstrated a pre-column CE-LIF assay to study the bioactivity of multiple enzymes, including hyaluronidase, elactase, and collagenase, in search of active cosmetic ingredients. By using a double detection besides LIF, another detection system, termed high-resolution mass spectrometry (HRMS), was connected in series to ensure the simultaneous analysis of three enzymatic reactions. All substrates and products were well defined in the less than 10 min run, with excellent limit of quantification LOQ (<5 nM) and good peak symmetry and efficiency, sufficient repeatability for intra-day and inter-day analysis (RSD < 4.5%) [106]. Prior to this work, the same group introduced a simple electroporation technique to destroy microalgae membranes to extract several amino acids in Dunaliella salina green algae, which were later analyzed by CE-LIF [107].

The compilation of using pre-capillary CE-LIF based in enzyme's kinetics, drug metabolism, or drug screening is listed in Table 2.

Table 2. Recent CE-LIF enzyme assays.

Enzyme	Substrate	Mode	Note	Ref.
neutrophil elastase	5-FAM-labeled peptides	on column	enzyme activity and inhibitor screening	[27]
protein kinase C	fluorescent-labeled peptide	off column	inhibitor screening	[59]
Sphingosine kinase	Fluorescein-labeled sphingosine	off column	kinase and phosphatase activity	[100]
AlkB	fluorescently labeled 15-nucleotide-long single-base methylated DNA substrate	off column	demethylation of DNA	[101]
histone deacetylase 1	5-carboxyfluorescein-labelled peptide	off column	inhibitor screening	[104]
Proteases	Fluorescence-labeled peptide	off column	proteolytic processing	[105]
Hyaluronidase, elastase and collagenase	FAM-peptides	off column	enzyme kinetics and plant substrate	[106]
Human neutrophil elastase	5-carboxyfluorescein (5-FAM) peptide	on column	enzyme kinetics, substrate study	[108]
alkaline phosphatase	MFP cyclohexylammonium	on column	enzyme kinetics and activity	[109]
glucose oxidase	2-[6-(4′-amino) phenoxy-3H-xanthen-3-on-9-yl] benzoic acid (APF)	on column	glucose determination, inhibitor screening	[110]
cholesterol oxidase	2-[6-(4′-amino) phenoxy-3H-xanthen-3-on-9-yl] benzoic acid (APF)	on column	cholesterol measurement	[111]
Lactate dehydrogenase	lactate	on column	enzymatic cycling reaction	[112]
recombinant human arylsulfatase	glycosaminoglycan	off column	enzyme kinetic, natural substrate	[113]
beta-glucosidase	Fluorescein mono-beta-D-glucopyranoside	off column	enzymatic activity	[114]
Calcineurin	Fluorescence-labeled 19-amino acid	off column	kinase activity	[115]
ATP	BODIPY FL EDA	off column	enzyme activity	[116]
Lysine decarboxylase	Lysine	off column	enzyme activity	[117]
oxygenases	AlkB	off column	inhibitor study	[118]
l-Asnase	FITC amino acids	off column	FITC amino acids	[118]
d-amino acid oxidase	d-amino acids	off column	enzyme activity	[118]
Signal peptidases	proprietary fluorescent-labeled substrate	off column	inhibitor screening	[119]
tyrosine kinase	fluorescence-labeled polypeptide substrate	off column	kinase activity, inhibitor screening	[120]
horseradish peroxidase	thyroxine, triiodothyronine, thyroid-stimulating hormone	on column	hormone study	[121]
recombinant GFP	thrombin	on column	enzyme activity	[122]
beta-galactosidase	resorufin-β-D-galactopyranoside	on column	Single molecule enzymology	[123]
protein farnesyltransferase	fluorescently labeled pentapeptide, farnesyl pyrophosphate	on column	inhibitor screening	[124]
alkaline phosphatase	AttoPhos	on column	enzyme inhibitor study	[125]
alkaline phosphatase	disodium phenyl phosphate	on column	enzyme catalysis	[126]

4.3. On-Column (In-Capillary) Enzymatic Assays

Bao and Regnier first pioneered the use of a capillary as a micro-reacting system for enzyme reactions [127]. This method was later termed electrophoretically mediated microanalysis (EMMA), or an on-column enzymatic assay. As suggested by the name, this technique integrates all reactions and separating steps into a single column, leading to further automation and extremely small consumption of enzymes, substrates, and cofactors. This approach is especially attractive where no sample workup is necessary and a high degree of miniaturization is achievable. In general, in-capillary assays encompass two modes to load enzyme and substrates, the continuous mode (long contact mode) and the plug–plug EMMA (short contact mode) [128]. In the continuous mode, the entire capillary is filled with either enzyme or substrate. Consequently, the second reactant is flushed through the channel as a plug. In the classical plug–plug EMMA, the enzyme and substrate are introduced as consecutive plugs, and the enzyme reaction takes place upon the application of an electric field. One drawback of EMMA is the incompatibility between the background electrolyte required for the separation and the enzyme reaction buffer. This shortcoming is not usually a big challenge, since in many cases background electrolytes consume similar buffers, which are used for sample preparation. However, to address this incompatibility, an additional plug of incubation buffer is injected into the system. This mode is named "partial filling mode" [129].

4.4. Application

The use of the EMMA approach has expanded in recent years for studies on enzyme activity and kinetics, inhibitor screening, substrate determination, and drug metabolism. In particular, the application of LIF detection has increased, especially in clinical or analogously oriented studies.

Fayad et al. reported an online EMMA-CE-LIF assay to evaluate the kinetic constant of a novel substrate of human neutrophil elastase (HNE), an enzyme responsible for skin aging and involved in the development of chronic obstructive pulmonary disease in non-small cell lung cancer progression [130,131]. Based on short-end injection, using transverse diffusion of laminar flow profiles (TDLFP) to mix the reactants, the analysis time was shortened to only 7 min [108]. The TDLFP mixing technique has also been used to study many enzyme activities [132–134]. HNE activity was assessed using both UV and LIF detection modes (Figure 6). The higher sensitivity obtained from LIF (almost three-fold in magnitude at a few nM LOD) is proof of the concept that LIF coupled with CE is economical and highly selective. The same group later developed another EMMA-CE-LIF-based method to study the inhibition of HNE [27]. For this, they utilized a complimentary technique termed microscale thermophoresis (MST), coupled with CE-LIF, to efficiently monitor the enzyme-inhibitor affinity. Benefiting from the previously published work, this assay to screen HNE inhibitors reached a high level of LOQ (~3 nM) with no pre-concentration steps required.

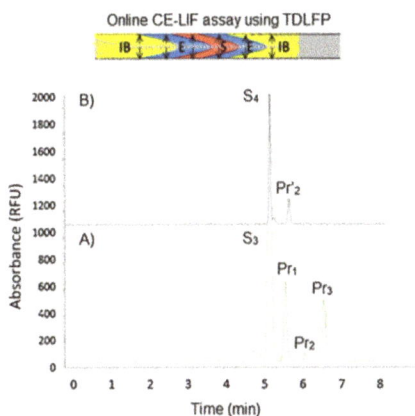

Figure 6. Electropherograms obtained by CE-LIF analysis for (**A**) S_3 or 5-FAM-Ala-Ala-Ala-Phe-Tyr-Asp-OH and (**B**) S_4 or 5-FAM-Arg-Glu-Ala-Val-Val-Tyr-OH hydrolysis by HNE. Reproduced with the permission of [108].

Coyle's group developed a method to study phosphatases from marine bacteria whose enzymatic functions are of importance to the mobilization, transformation, and turnover of compounds in aquatic environments [109]. Based on the precedence of previous work [135], they determined the alkaline phosphatase's kinetics in four marine proteobacteria isolates, using a fluorescently labeled substrate 3-o-methylfluorescein phosphate cyclohexylammonium salt MFP. Without significant modifications, the CE-LIF online approach provided more information about enzyme diversity and could be applied to study the phosphatases in various organisms. However, challenges remain for the method, including the identification and characterization of enzymes which have the same electrophoretic mobility.

Diabetes is one of the most common metabolic disorders worldwide, characterized by high blood sugar levels in patient serum [136,137]. Due to its lethal complications, many research groups have been extensively seeking to develop fast, accurate, and reliable methods to read and/or monitor glucose. Recently, Guan et al. introduced an ultrasensitive analysis of glucose in serum, using a CE-LIF on-column enzymatic assay [110]. The reaction between glucose oxidase (GOx) and horseradish peroxidase (HRP) released hydrogen peroxide (H2O2), leading to the activation of a fluorogenic reagent named 2-[6-(4'-amino) phenoxy-3H-xanthen-3-on-9-yl] benzoic acid APF to form a highly fluorescent product, which was later electrophoretically separated from unreacted APF by the LIF detection. Their proposed method allowed the detection of glucose in real samples down to 10 nM, and RSD values lower than 3.5%.

In addition, other components of biological fluids could be targets using the EMMA-CE-LIF platform. The same group reported this method in the use of determining total cholesterol in human plasma [111].

Nicotinamide adenine dinucleotide (NAD+) and its reduced form NADH act as two coenzymes that are involved in cellular energy metabolism, including glycolysis and oxidative phosphorylation. The conversion of the two enzymes, or the NADH/NAD+, ratio reflects the cellular metabolic status, thus responding to environmental stimuli. Xie and colleagues developed an in-capillary assay to detect NAD+ and NADH contents of a single cell, by coupling an enzymatic cycling reaction with CE-LIF [112]. NAD+ is reduced to NADH in one enzymatic reaction, and in return NADH is oxidized to NAD+ with the production of a fluorescent product [138,139]. As the cycle goes on, with the formation of a fluorescent product, resorufin, the accumulation rate is therefore correlated to the NAD+/NADH ratio in the system. Xie reported the LOD for NAD+ using on-line capillary assays being as low as 0.2 nM, with the reproducibility (RSD) of 5%, which is much more sensitive than that of concurrent methods available for the determination of NAD+ and NADH in cells and tissues, such as fluorescence imaging, enzymatic assay, or HPLC.

The recent EMMA-CE-LIF methods are listed in Table 2.

4.5. Chip-Based Enzyme Assays and Application in CE-LIF Systems

In general, the application of microchip devices in capillary electrophoresis has been targeted to acquire the miniaturization and portability features. To join in LIF mode, a significant reduction in the size of the auxiliary apparatus is necessary to achieve such a sophisticated system. Initially, since the laser module is bulk equipment, microchips have acted as a sample preparation tool in need of pre-concentration or purification. However, several research groups have been seeking methods to integrate the whole fluorescence detector inside the microchips, a new concept known as "lab-on-a-chip". The most common strategy is to embed optical fibers inside the micro devices to bring the light in and out of the detection unit. Various designs have been introduced with the optical fiber inserted onto the microchip. Among them, optical micro lenses have become a trend, using different materials and techniques. For examples, polymer micro lenses have been utilized to focus light inside the microfluidic channel and increase the excitation of the fluorescent tags. To apply this strategy and aim for high-throughput in design, Guestchow's group fabricated polydimethylsiloxane (PDMS) on a chip device [140]. By taking advantage of the intrinsic properties of glass (hydrophilic) and PDMS (hydrophobic), the extraction of droplets from the segmented flow was simplified. The segmented flow sample streams were coupled with hybrid PDMS-chip glass and rapidly analyzed by microchip electrophoresis with a dual beam laser-induced system. A total of 160 test compounds as potential inhibitors of protein kinase A were subsequently screened and each sample generated two droplets, thus allowing approximately six injections per sample. Including controls (negative and positive), a total of 168 samples were analyzed in approximately 12 min. Of the screened compounds, 25 potential enzyme inhibitors were identified, with the IC50 of two inhibitors calculated.

Microchips in CE usually use a very short separation distance. Therefore, it is crucial to effectively deliver narrow and reproducible sample plugs into the separation channels. Optically gated sample introduction is an alternative injection method from the commonly used T-type injection. In this injection mode, fluorescently labeled analyte is electrophoresed through the separation section. A laser is split into two beams and focused on the two points of the separation section. One beam (the probe beam) has less laser power and is utilized for LIF detection. The other beam (the gating beam) has higher laser power and is used to perform the sample introduction by time-discriminated photobleaching of the sample. For instance, Gong et al. introduced a microfluidic device-based enzyme assay using beta-glucuronidase as the enzyme and a fluorescein di-(beta-D-glucuronidase) as the substrate model [141]. Their device was coupled with a dual-beam LIF detection, where the fluorescent signals were recorded with two independent LIF channels capable of being aligned to various positions on the device. Consequently, the enzymatic product concentrations in the manner of time incubation were measured, the Michalis constant was determined, and inhibition assays were carried using a competitive inhibitor to calculate the IC50 values of promising inhibitors.

In the study of Belder's group, they demonstrated a rapid CE-chip-based assay to investigate the deacetylation of acetyl-lysine residues by sirtuins (SIRTs) enzymes [142]. Their microchip design consisted of a microfluidic separation structure integrated with a serpentine micromixer. 9-fluorenylmethoxycarbonyl (Fmoc)-labeled tetrapeptide derived from p53 was used as the SIRT substrate. The substrate and enzyme were mixed in the reaction channel and subsequently transferred to the separation channel by an electrophoretic pinched injection. The SIRT inhibitors were screened, with the results in alignment with the literature. The total time, including incubation, chemical reaction, and analysis, was about 30 min, which was shorter than the previously published techniques [143].

Besides reaction in solution, enzymes can also be immobilized to solid supports, known as immobilized enzyme reactors (IMERs), which are incorporated into CE-chips to further utilize the EMMA technique in a miniaturized system. Reviews were published using IMERs in proteomics [144] or microfluidic process reactors [145] and field-analysis [84]. Microchip CE has employed IMERs in its offline approach, in combination with the LIF detection mode. For example, Qiao published a series of microchip CE-LIF assays to study the enzyme activities of L-asparaginase [146,147]. Their results demonstrated a promising therapeutic protocol for acute lymphoblastic leukemia.

5. Conclusions and Perspective Outlook

This review discussed the application of CE-LIF as a tool to develop immunoassays and enzyme assays and its application in the last decade. Due to the increasing demand in biological, clinical, and pharmaceutical research, the use of CE-LIF has been expanding rapidly, because of its ultra-sensitivity and selectivity. The instruments and method are continually changing to improve the efficiency of the analysis, as well as bridging the application of CE to broader spectrums of analysis. Scientists are extensively seeking to modify the system by introducing new materials for capillary column modification, or analyte derivatization to maximally enhance the analytical performance. Emerging as a new trend, nanoparticles (NPs) with a size of less than 100 nm have attracted much attention for a widespread range of applications, due to their unique physicochemical characteristics and facile surface modification. Among various NPs, employment of magnetic beads as solid support for immunoextraction has been seen in many reports [148–152]. The number of studies and applications of NPs as independent surrogates or in coupling with the microfluidic process in immunoaffinity and the enzymatic assays is continuously increasing, with numerous papers dealing with NPs in the CE system.

To improve the performance of complex biological and biomedical samples, multidimensional separation has become a tendency, where two or more orthogonal displacement mechanisms are combined in CE systems. Especially with regards to fluorescence detection, multiple dimensions have helped overcome the matrix interferences, thus significantly improving the sensitivity. Together with the combination of microfluidic devices, CE-LIF has enhanced efficiency, as well as allowing the multiplexed separation of analytes. An integrated 2D, 3D, or more dimensional system in the future will make CE separation a portable and universal tool of practice in medicine, pharmaceuticals, environmental science, and food science.

LIF detection has been well established for decades in the capillary electrophoresis field. However, in an attempt to reduce the cost of expensive laser modules and avoid stability issues in the baseline (especially in the UV range), researchers have been looking for alternative light sources for LIF. CE-LEDIF has been a replacement candidate, which was discussed briefly earlier due to the various emission wavelengths that LED sources can offer. In terms of detection limit, it is still controversial compared to LIF, however, in the near future, LEDIF can possibly replace LIF detection.

We discussed herein the CE-LIF immunoassays and pointed out the advantages, disadvantages, and points to consider of each format. For immunoassays, competitive homogenous binding assays are still the most popular of the CE-LIF based immunoassay methods. We also included a discussion of the enzymatic reaction. Both kinds of reaction have been applied successfully to capillary electrophoresis. Of enzyme assays, offline methods are easy to perform and manage. With a combination of EMMA, IMERs, and multidimensional channels, we attempted to bring to light the immunoassay and enzyme assay point of view to the analytical scientists, and highlight the opportunity to harness this powerful technique within analytical chemistry. Table 3 lists the common advantages and disadvantages offered by CE-LIF for these bioassays.

Intriguingly, chip devices have inevitably proved their capacity to incorporate with CE to become a fast, high throughput, and automated miniaturization system. We covered the application of microfluidic devices in both categories to showcase the volatility and ease of use upon the demanding and challenging requirements of industrial purposes. At its core, CE-LIF can become a prognostic or diagnostic tool in multiple clinical fields.

Table 3. CE-LIF based assay advantages and disadvantages.

Advantages	Disadvantages
-High sensitivity, high speed -High reproducibility -Small number of samples required -Minimal preparation time -Easy automation, high throughput for profiling of complex biological samples -Possible on-column concentration -Highly efficient multidimensional separation	-Derivatization required -Instability of the laser power -Excitation range is limited -No standardized method

Funding: This research was funded by Korea Institute of Science and Technology institutional program, grant number 2E29290.

Acknowledgments: This research was supported by the Korea Institute of Science and Technology institutional program (2E29290).

Conflicts of Interest: The authors declare no conflict of interest.

References

1. Jorgenson, J.W.; Lukacs, K.D. Zone electrophoresis in open-tubular glass capillaries. *Anal. Chem.* **1981**, *53*, 1298–1302. [CrossRef]
2. Jorgenson, J.W.; Lukacs, K.D. Capillary zone electrophoresis. *Science* **1983**, *222*, 266–272. [CrossRef] [PubMed]
3. Aid, T.; Paist, L.; Lopp, M.; Kaljurand, M.; Vaher, M. An optimized capillary electrophoresis method for the simultaneous analysis of biomass degradation products in ionic liquid containing samples. *J. Chromatogr. B* **2016**, *1447*, 141–147. [CrossRef]

4. Yu, J.; Aboshora, W.; Zhang, S.; Zhang, L. Direct UV determination of Amadori compounds using ligand-exchange and sweeping capillary electrophoresis. *Anal. Bioanal. Chem.* **2016**, *408*, 1657–1666. [CrossRef]

5. Bucsella, B.; Fornage, A.; Denmat, C.L.; Kalman, F. Nucleotide and Nucleotide Sugar Analysis in Cell Extracts by Capillary Electrophoresis. *Chimia* **2016**, *70*, 732–735. [CrossRef] [PubMed]

6. Hiltunen, S.; Sirén, H.; Heiskanen, I.; Backfolk, K. Capillary electrophoretic profiling of wood-based oligosaccharides. *Cellulose* **2016**, *23*, 3331–3340. [CrossRef]

7. Wenz, C.; Barbas, C.; Lopez-Gonzalvez, A.; Garcia, A.; Benavente, F.; Sanz-Nebot, V.; Blanc, T.; Freckleton, G.; Britz-McKibbin, P.; Shanmuganathan, M.; et al. Interlaboratory study to evaluate the robustness of capillary electrophoresis-mass spectrometry for peptide mapping. *J. Sep. Sci.* **2015**, *38*, 3262–3270. [CrossRef] [PubMed]

8. Guo, X.; Fillmore, T.L.; Gao, Y.; Tang, K. Capillary Electrophoresis-Nanoelectrospray Ionization-Selected Reaction Monitoring Mass Spectrometry via a True Sheathless Metal-Coated Emitter Interface for Robust and High-Sensitivity Sample Quantification. *Anal. Chem.* **2016**, *88*, 4418–4425. [CrossRef]

9. Han, M.; Rock, B.M.; Pearson, J.T.; Rock, D.A. Intact mass analysis of monoclonal antibodies by capillary electrophoresis-Mass spectrometry. *J. Chromatogr. B Analyt. Technol. Biomed. Life Sci.* **2016**, *1011*, 24–32. [CrossRef] [PubMed]

10. Negri, P.; Sarver, S.A.; Schiavone, N.M.; Dovichi, N.J.; Schultz, Z.D. Online SERS detection and characterization of eight biologically-active peptides separated by capillary zone electrophoresis. *Analyst* **2015**, *140*, 1516–1522. [CrossRef]

11. Negri, P.; Flaherty, R.J.; Dada, O.O.; Schultz, Z.D. Ultrasensitive online SERS detection of structural isomers separated by capillary zone electrophoresis. *Chem. Com.* **2014**, *50*, 2707–2710. [CrossRef] [PubMed]

12. Fan, Y.; Scriba, G.K. Advances in capillary electrophoretic enzyme assays. *J. Pharm. Biomed. Anal.* **2010**, *15*, 1076–1090. [CrossRef] [PubMed]

13. Hai, X.; Yang, B.F.; Schepdael, A.V. Recent developments and applications of EMMA in enzymatic and derivatization reactions. *Electrophoresis* **2012**, *33*, 211–227. [CrossRef] [PubMed]

14. Zhang, J.; Hoogmartens, J.; Van Schepdael, A. Advances in capillary electrophoretically mediated microanalysis: An update. *Electrophoresis* **2006**, *27*, 35–43. [CrossRef]

15. Bonin, L.; Aupiais, J.; Kerbaa, M.; Moisy, P.; Topin, S.; Siberchicot, B. Revisiting actinide–DTPA complexes in aqueous solution by CE-ICPMS and ab initio molecular dynamics. *RSC Adv.* **2016**, *6*, 62729–62741. [CrossRef]

16. Mai, T.D.; Le, M.D.; Saiz, J.; Duong, H.A.; Koenka, I.J.; Pham, H.V.; Hauser, P.C. Triple-channel portable capillary electrophoresis instrument with individual background electrolytes for the concurrent separations of anionic and cationic species. *Anal. Chim. Acta* **2016**, *911*, 121–128. [CrossRef] [PubMed]

17. Huhner, J.; Jooss, K.; Neususs, C. Interference-free mass spectrometric detection of capillary isoelectric focused proteins, including charge variants of a model monoclonal antibody. *Electrophoresis* **2017**, *38*, 914–921. [CrossRef] [PubMed]

18. Kanoatov, M.; Krylov, S.N. Analysis of DNA in Phosphate Buffered Saline Using Kinetic Capillary Electrophoresis. *Anal. Chem.* **2016**, *88*, 7421–7428. [CrossRef]

19. Tohala, L.; Oukacine, F.; Ravelet, C.; Peyrin, E. Sequence requirements of oligonucleotide chiral selectors for the capillary electrophoresis resolution of low-affinity DNA binders. *Electrophoresis* **2017**, *38*, 1383–1390. [CrossRef]

20. Mofaddel, N.; Fourmentin, S.; Guillen, F.; Landy, D.; Gouhier, G. Ionic liquids and cyclodextrin inclusion complexes: Limitation of the affinity capillary electrophoresis technique. *Anal. Bioanal. Chem.* **2016**, *408*, 8211–8220. [CrossRef]

21. Van Tricht, E.; Geurink, L.; Backus, H.; Germano, M.; Somsen, G.W.; Sanger-van de Griend, C.E. One single, fast and robust capillary electrophoresis method for the direct quantification of intact adenovirus particles in upstream and downstream processing samples. *Talanta* **2017**, *166*, 8–14. [CrossRef]

22. Rodrigues, K.T.; Mekahli, D.; Tavares, M.F.; Van Schepdael, A. Development and validation of a CE-MS method for the targeted assessment of amino acids in urine. *Electrophoresis* **2016**, *37*, 1039–1047. [CrossRef]

23. Suba, D.; Urbanyi, Z.; Salgo, A. Method development and qualification of capillary zone electrophoresis for investigation of therapeutic monoclonal antibody quality. *J. Chromatogr. B Analyt. Technol. Biomed. Life Sci.* **2016**, *1032*, 224–229. [CrossRef]

24. Hamm, M.; Wang, F.; Rustandi, R.R. Development of a capillary zone electrophoresis method for dose determination in a tetravalent dengue vaccine candidate. *Electrophoresis* **2015**, *36*, 2687–2694. [CrossRef]
25. Jooss, K.; Huhner, J.; Kiessig, S.; Moritz, B.; Neususs, C. Two-dimensional capillary zone electrophoresis-mass spectrometry for the characterization of intact monoclonal antibody charge variants, including deamidation products. *Anal. Bioanal. Chem.* **2017**, *409*, 6057–6067. [CrossRef] [PubMed]
26. Xiao, L.; Liu, S.; Lin, L.; Yao, S. A CIEF-LIF method for simultaneous analysis of multiple protein kinases and screening of inhibitors. *Electrophoresis* **2016**, *37*, 2075–2082. [CrossRef]
27. Syntia, F.; Nehmé, R.; Claude, B.; Morin, P. Human neutrophil elastase inhibition studied by capillary electrophoresis with laser induced fluorescence detection and microscale thermophoresis. *J. Chromatogr. A* **2016**, *1431*, 215–223. [CrossRef]
28. Stephen, T.K.; Guillemette, K.L.; Green, T.K. Analysis of Trinitrophenylated Adenosine and Inosine by Capillary Electrophoresis and gamma-Cyclodextrin-Enhanced Fluorescence Detection. *Anal. Chem.* **2016**, *88*, 7777–7785. [CrossRef]
29. Goedecke, S.; Muhlisch, J.; Hempel, G.; Fruhwald, M.C.; Wunsch, B. Quantitative analysis of DNA methylation in the promoter region of the methylguanine-O(6)-DNA-methyltransferase gene by COBRA and subsequent native capillary gel electrophoresis. *Electrophoresis* **2015**, *36*, 2939–2950. [CrossRef]
30. Meininger, M.; Stepath, M.; Hennig, R.; Cajic, S.; Rapp, E.; Rotering, H.; Wolff, M.W.; Reichl, U. Sialic acid-specific affinity chromatography for the separation of erythropoietin glycoforms using serotonin as a ligand. *J. Chromatogr. B Analyt. Technol. Biomed. Life Sci.* **2016**, *1012–1013*, 193–203. [CrossRef]
31. Albrecht, S.; Schols, H.A.; Klarenbeek, B.; Voragen, A.G.J.; Gruppen, H. Introducing Capillary Electrophoresis with Laser-Induced Fluorescence (CE–LIF) as a Potential Analysis and Quantification Tool for Galactooligosaccharides Extracted from Complex Food Matrices. *J. Agric. Food Chem.* **2010**, *58*, 2787–2794. [CrossRef] [PubMed]
32. Kovacs, Z.; Szarka, M.; Szigeti, M.; Guttman, A. Separation window dependent multiple injection (SWDMI) for large scale analysis of therapeutic antibody N-glycans. *J. Pharm. Biomed. Anal.* **2016**, *128*, 367–370. [CrossRef] [PubMed]
33. Moser, A.C.; Hage, D.S. Capillary Electrophoresis-Based Immunoassays: Principles & Quantitative Applications. *Electrophoresis* **2008**, *29*, 3279–3295. [PubMed]
34. Amundsen, L.K.; Siren, H. Immunoaffinity CE in clinical analysis of body fluids and tissues. *Electrophoresis* **2007**, *28*, 99–113. [CrossRef]
35. Schmalzing, D.; Nashabeh, W. Capillary electrophoresis based immunoassays: A critical review. *Electrophoresis* **1997**, *18*, 2184–2193. [CrossRef] [PubMed]
36. Yeung, W.S.; Luo, G.A.; Wang, Q.G.; Ou, J.P. Capillary electrophoresis-based immunoassay. *J. Chromatogr. B Analyt. Technol. Biomed. Life Sci.* **2003**, *25*, 217–228. [CrossRef]
37. Yeung, E.S.G.; Guzman, N.A. *Capillary Electrophoresis Technology*; Taylor & Francis Inc.: Broken Sound Parkway, NW, USA, 1993; pp. 592–597.
38. Wu, C.; Sun, Y.; Wang, Y.; Duan, W.; Hu, J.; Zhou, L.; Pu, Q. 7-(Diethylamino)coumarin-3-carboxylic acid as derivatization reagent for 405 nm laser-induced fluorescence detection: A case study for the analysis of sulfonamides by capillary electrophoresis. *Talanta* **2019**, *201*, 16–22. [CrossRef]
39. Kuo, J.S.; Kuyper, C.L.; Allen, P.B.; Fiorini, G.S.; Chiu, D.T. High-power blue/UV light-emitting diodes as excitation sources for sensitive detection. *Electrophoresis* **2004**, *25*, 3796–3804. [CrossRef] [PubMed]
40. Williams, D.C.; Soper, S.A. Ultrasensitive Near-IR Fluorescence Detection for Capillary Gel Electrophoresis and DNA Sequencing Applications. *Anal. Chem.* **1995**, *67*, 3427–3432. [CrossRef]
41. McWhorter, S.; Soper, S.A. Near-infrared laser-induced fluorescence detection in capillary electrophoresis. *Electrophoresis* **2000**, *21*, 1267–1280. [CrossRef]
42. Aboul-Enein, H.Y.; Ali, I. Capillary electrophoresis. In *Analytical Instrumentation Handbook*; Taylor & Francis Inc.: Broken Sound Parkway, NW, USA, 2004; pp. 803–826.
43. Kao, Y.-Y.; Liu, K.-T.; Huang, M.-F.; Chiu, T.-C.; Chang, H.-T. Analysis of amino acids and biogenic amines in breast cancer cells by capillary electrophoresis using polymer solutions containing sodium dodecyl sulfate. *J. Chromatogr. A* **2010**, *1217*, 582–587. [CrossRef] [PubMed]
44. Chang, P.L.; Chiu, T.C.; Wang, T.E.; Hu, K.C.; Tsai, Y.H.; Hu, C.C.; Bair, M.J.; Chang, H.T. Quantitation of branched-chain amino acids in ascites by capillary electrophoresis with light-emitting diode-induced fluorescence detection. *Electrophoresis* **2011**, *32*, 1080–1083. [CrossRef] [PubMed]

45. Lin, K.-C.; Hsieh, M.-M.; Chang, C.-W.; Lin, E.-P.; Wu, T.-H. Stacking and separation of aspartic acid enantiomers under discontinuous system by capillary electrophoresis with light-emitting diode-induced fluorescence detection. *Talanta* **2010**, *82*, 1912–1918. [CrossRef] [PubMed]
46. Yang, F.; Li, X.-c.; Zhang, W.; Pan, J.-b.; Chen, Z.-g. A facile light-emitting-diode induced fluorescence detector coupled to an integrated microfluidic device for microchip electrophoresis. *Talanta* **2011**, *84*, 1099–1106. [CrossRef] [PubMed]
47. Grochocki, W.; Buszewska-Forajta, M.; Macioszek, S.; Markuszewski, M.J. Determination of urinary pterins by capillary electrophoresis coupled with LED-induced fluorescence detector. *Molecules* **2019**, *24*, 24. [CrossRef]
48. An, J.Y.; Azizov, S.; Kumar, A.P.; Lee, Y.I. Quantitative Analysis of Artificial Sweeteners by Capillary Electrophoresis with a Dual-Capillary Design of Molecularly Imprinted Solid-Phase Extractor. *Bull. Korean Chem. Soc.* **2018**, *39*, 1315–1319. [CrossRef]
49. Ji, H.; Zhang, X.; Yang, F.; Wang, J.; Yuan, H.; Xiao, D. Sensitive determination of l-hydroxyproline in dairy products by capillary electrophoresis with in-capillary optical fiber light-emitting diode-induced fluorescence detection. *Anal. Methods* **2018**, *10*, 2211–2216. [CrossRef]
50. Hühner, J.; Ingles-Prieto, A.; Neusüß, C.; Lämmerhofer, M.; Janovjak, H. Quantification of riboflavin, flavin mononucleotide, and flavin adenine dinucleotide in mammalian model cells by CE with LED-induced fluorescence detection. *Electrophoresis* **2015**, *36*, 518–525. [CrossRef]
51. Le Potier, I.; Boutonnet, A.; Ecochard, V.; Couderc, F. Chemical and instrumental approaches for capillary electrophoresis (CE)-fluorescence analysis of proteins. *Methods Biochem. Anal.* **2016**, *1466*, 1–10.
52. Yang, X.; Yan, W.; Bai, H.; Lv, H.; Liu, Z. A scanning laser induced fluorescence detection system for capillary electrophoresis microchip based on optical fiber. *Optik* **2012**, *123*, 2126–2130. [CrossRef]
53. Dada, O.O.; Ramsay, L.M.; Dickerson, J.A.; Cermak, N.; Jiang, R.; Zhu, C.; Dovichi, N.J. Capillary array isoelectric focusing with laser-induced fluorescence detection: Milli-pH unit resolution and yoctomole mass detection limits in a 32-channel system. *Anal. Bioanal. Chem.* **2010**, *397*, 3305–3310. [CrossRef] [PubMed]
54. Melanson, J.E.; Lucy, C.A. Violet (405 nm) diode laser for laser induced fluorescence detection in capillary electrophoresis. *Analyst* **2000**, *125*, 1049–1052. [CrossRef]
55. Harrison, S.; Geppetti, P. Substance p. *Int. J. Biochem. Cell Biol.* **2001**, *33*, 555–576. [CrossRef]
56. Michels, D.A.; Hu, S.; Schoenherr, R.M.; Eggertson, M.J.; Dovichi, N.J. Fully Automated Two-dimensional Capillary Electrophoresis for High Sensitivity Protein Analysis. *Mol. Cell Proteom.* **2002**, *1*, 69–74. [CrossRef]
57. Zhang, Z.; Carpenter, E.; Puyan, X.; Dovichi, N.J. Manipulation of protein fingerprints during on-column fluorescent labeling: Protein fingerprinting of six Staphylococcus species by capillary electrophoresis. *Electrophoresis* **2001**, *22*, 1127–1132. [CrossRef]
58. Novatchev, N.; Holzgrabe, U. Evaluation of the impurity profile of amino acids by means of CE. *J. Pharm. Biomed. Anal.* **2001**, *26*, 779–789. [CrossRef]
59. Nguyen, B.T.; Park, M.; Pyun, J.C.; Yoo, Y.S.; Kang, M.J. Efficient PKC inhibitor screening achieved using a quantitative CE-LIF assay. *Electrophoresis* **2016**, *37*, 3146–3153. [CrossRef] [PubMed]
60. Nguyen, B.T.; Park, M.; Yoo, Y.S.; Kang, M.J. Capillary electrophoresis-laser-induced fluorescence (CE-LIF)-based immunoassay for quantifying antibodies against cyclic citrullinated peptides. *Analyst* **2018**, *143*, 3141–3147. [CrossRef] [PubMed]
61. Shimura, K.; Kamiya, K.-i.; Matsumoto, H.; Kasai, K.-i. Fluorescence-Labeled Peptide pI Markers for Capillary Isoelectric Focusing. *Anal. Chem.* **2002**, *74*, 1046–1053. [CrossRef]
62. Korchane, S.; Pallandre, A.; Przybylski, C.; Pous, C.; Gonnet, F.; Taverna, M.; Daniel, R.; Le Potier, I. Derivatization strategies for CE-LIF analysis of biomarkers: Toward a clinical diagnostic of familial transthyretin amyloidosis. *Electrophoresis* **2014**, *35*, 1050–1059. [CrossRef]
63. Moody, E.D.; Viskari, P.J.; Colyer, C.L. Non-covalent labeling of human serum albumin with indocyanine green: A study by capillary electrophoresis with diode laser-induced fluorescence detection. *J. Chromatogr. B Biomed. Sci. Appl.* **1999**, *729*, 55–64. [CrossRef]
64. Liu, C.; Deng, Q.; Fang, G.; Dang, M.; Wang, S. Capillary electrochromatography immunoassay for alpha-fetoprotein based on poly(guanidinium ionic liquid) monolithic material. *Anal. Biochem.* **2017**, *530*, 50–56. [CrossRef] [PubMed]
65. Wang, X.; Song, Y.; Song, M.; Wang, Z.; Li, T.; Wang, H. Fluorescence Polarization Combined Capillary Electrophoresis Immunoassay for the Sensitive Detection of Genomic DNA Methylation. *Anal. Chem.* **2009**, *81*, 7885–7891. [CrossRef]

66. Hao, L.; Bai, Y.; Wang, H.; Zhao, Q. Affinity capillary electrophoresis with laser induced fluorescence detection for thrombin analysis using nuclease-resistant RNA aptamers. *J. Chromatogr. A* **2016**, *1476*, 124–129. [CrossRef] [PubMed]

67. Song, M.; Zhang, Y.; Li, T.; Wang, Z.; Yin, J.; Wang, H. Highly sensitive detection of human thrombin in serum by affinity capillary electrophoresis/laser-induced fluorescence polarization using aptamers as probes. *J. Chromatogr. A* **2009**, *1216*, 873–878. [CrossRef]

68. Yi, L.; Wang, X.; Bethge, L.; Klussmann, S.; Roper, M.G. Noncompetitive affinity assays of glucagon and amylin using mirror-image aptamers as affinity probes. *Analyst* **2016**, *141*, 1939–1946. [CrossRef]

69. Zhang, H.; Li, X.-F.; Le, X.C. Tunable Aptamer Capillary Electrophoresis and Its Application to Protein Analysis. *JACS* **2008**, *130*, 34–35. [CrossRef] [PubMed]

70. Shen, R.; Guo, L.; Zhang, Z.; Meng, Q.; Xie, J. Highly sensitive determination of recombinant human erythropoietin-α in aptamer-based affinity probe capillary electrophoresis with laser-induced fluorescence detection. *J. Chromatogr. A* **2010**, *1217*, 5635–5641. [CrossRef]

71. Perrier, S.; Zhu, Z.; Fiore, E.; Ravelet, C.; Guieu, V.; Peyrin, E. Capillary Gel Electrophoresis-Coupled Aptamer Enzymatic Cleavage Protection Strategy for the Simultaneous Detection of Multiple Small Analytes. *Anal. Chem.* **2014**, *86*, 4233–4240. [CrossRef]

72. Giovannoli, C.; Baggiani, C.; Passini, C.; Biagioli, F.; Anfossi, L.; Giraudi, G. A rational route to the development of a competitive capillary electrophoresis immunoassay: Assessment of the variables affecting the performances of a competitive capillary electrophoresis immunoassay for human serum albumin. *Talanta* **2012**, *94*, 65–69. [CrossRef]

73. Liu, C.; Feng, X.; Qian, H.; Fang, G.; Wang, S. Determination of Norfloxacin in Food by Capillary Electrophoresis Immunoassay with Laser-Induced Fluorescence Detector. *Food Anal. Methods* **2015**, *8*, 596–603. [CrossRef]

74. Zhang, C.; Ma, G.; Fang, G.; Zhang, Y.; Wang, S. Development of a Capillary Electrophoresis-Based Immunoassay with Laser-Induced Fluorescence for the Detection of Carbaryl in Rice Samples. *J. Agric. Food Chem.* **2008**, *56*, 8832–8837. [CrossRef] [PubMed]

75. Zhang, C.; Wang, S.; Fang, G.; Zhang, Y.; Jiang, L. Competitive immunoassay by capillary electrophoresis with laser-induced fluorescence for the trace detection of chloramphenicol in animal-derived foods. *Electrophoresis* **2008**, *29*, 3422–3428. [CrossRef]

76. Zhang, H.; Li, X.F.; Le, X.C. Differentiation and detection of PDGF isomers and their receptors by tunable aptamer capillary electrophoresis. *Anal. Chem.* **2009**, *81*, 7795–7800. [CrossRef]

77. Chen, H.-X.; Zhang, X.-X. Antibody development to testosterone and its application in capillary electrophoresis-based immunoassay. *Electrophoresi* **2008**, *29*, 3406–3413. [CrossRef]

78. Yu, X.; He, Y.; Jiang, J.; Cui, H. A competitive immunoassay for sensitive detection of small molecules chloramphenicol based on luminol functionalized silver nanoprobe. *Anal. Chim. Acta* **2014**, *812*, 236–242. [CrossRef]

79. Lomasney, A.R.; Guillo, C.; Sidebottom, A.M.; Roper, M.G. Optimization of capillary electrophoresis conditions for a glucagon competitive immunoassay using response surface methodology. *Anal. Biol. Anal. Chem.* **2009**, *394*, 313–319. [CrossRef]

80. Giovannoli, C.; Anfossi, L.; Baggiani, C.; Giraudi, G. Binding properties of a monoclonal antibody against the Cry1Ab from Bacillus Thuringensis for the development of a capillary electrophoresis competitive immunoassay. *Anal. Bioanal. Chem.* **2008**, *392*, 385–393. [CrossRef]

81. Zhu, Z.; Ravelet, C.; Perrier, S.; Guieu, V.; Roy, B.; Perigaud, C.; Peyrin, E. Multiplexed Detection of Small Analytes by Structure-Switching Aptamer-Based Capillary Electrophoresis. *Anal. Chem.* **2010**, *82*, 4613–4620. [CrossRef]

82. Harrison, D.J.; Manz, A.; Fan, Z.; Luedi, H.; Widmer, H.M. Capillary electrophoresis and sample injection systems integrated on a planar glass chip. *Anal. Chem.* **1992**, *64*, 1926–1932. [CrossRef]

83. Wuethrich, A.; Quirino, J.P. A decade of microchip electrophoresis for clinical diagnostics—A review of 2008–2017. *Anal. Chim. Acta* **2019**, *1045*, 42–66. [CrossRef]

84. Zhang, M.; Phung, S.C.; Smejkal, P.; Guijt, R.M.; Breadmore, M.C. Recent trends in capillary and micro-chip electrophoretic instrumentation for field-analysis. *Trends Environ. Anal.* **2018**, *18*, 1–10. [CrossRef]

85. Kašička, V. Recent developments in capillary and microchip electroseparations of peptides (2013–middle 2015). *Electrophoresis* **2016**, *37*, 162–188. [CrossRef]

86. Sonker, M.; Sahore, V.; Woolley, A.T. Recent advances in microfluidic sample preparation and separation techniques for molecular biomarker analysis: A critical review. *Anal. Chim. Acta* **2017**, *986*, 1–11. [CrossRef]

87. Zare, R.N.; Kim, S. Microfluidic Platforms for Single-Cell Analysis. *Annu. Rev. Biomed. Eng.* **2010**, *12*, 187–201. [CrossRef] [PubMed]

88. Hage, D.S. Immunoassays. *Anal. Chem.* **1999**, *71*, 294r–304r. [CrossRef]

89. Giovannoli, C.; Anfossi, L.; Baggiani, C.; Giraudi, G. novel approach for a non competitive capillary electrophoresis immunoassay with laser-induced fluorescence detection for the determination of human serum albumin. *J. Chromatogr. A* **2007**, *6*, 187–192. [CrossRef] [PubMed]

90. Roper, M.G.; Shackman, J.G.; Dahlgren, G.M.; Kennedy, R.T. Microfluidic chip for continuous monitoring of hormone secretion from live cells using an electrophoresis-based immunoassay. *Anal. Chem.* **2003**, *75*, 4711–4717. [CrossRef]

91. Phillips, T.M.; Wellner, E.F. Chip-based immunoaffinity CE: Application to the measurement of brain-derived neurotrophic factor in skin biopsies. *Electrophoresis* **2009**, *30*, 2307–2312. [CrossRef]

92. Phillips, T.M.; Wellner, E. Detection of cerebral spinal fluid-associated chemokines in birth traumatized premature babies by chip-based immunoaffinity CE. *Electrophoresis* **2013**, *34*, 1530–1538. [CrossRef] [PubMed]

93. Phillips, T.M.; Wellner, E.F. Analysis of Inflammatory Mediators in Newborn Dried Blood Spot Samples by Chip-Based Immunoaffinity Capillary Electrophoresis. In *Clinical Applications of Capillary Electrophoresis: Methods and Protocols*; Phillips, T.M., Ed.; Springer: New York, NY, USA, 2019; pp. 185–198.

94. Shi, M.; Zhao, S.; Huang, Y.; Liu, Y.-M.; Ye, F. Microchip fluorescence-enhanced immunoaasay for simultaneous quantification of multiple tumor markers. *J. Chromatogr. B* **2011**, *879*, 2840–2844. [CrossRef] [PubMed]

95. Banke, N.; Hansen, K.; Diers, I. Detection of enzyme activity in fractions collected from free solution capillary electrophoresis of complex samples. *J. Chromatogr. A* **1991**, *559*, 325–335. [CrossRef]

96. Nguyen, H.T.; Waldrop, G.L.; Gilman, D.L. Capillary electrophoretic assay of human acetyl-coenzyme A carboxylase 2. *Electrophoresis* **2019**, *0*, 1–7. [CrossRef]

97. Bryatt, S.K.; Waldrop, G.L.; Gilman, D.L. A Capillary Electrophoretic Assay for Acetyl CoA Carboxylase. *Anal. Biochem.* **2013**, *437*, 32–38.

98. Chen, C.; Bonisch, D.; Penzis, R.; Winckler, T.; Scriba, G.K.E. Capillary Electrophoresis-Based Enzyme Assay for Nicotinamide N-Methyltransferase. *Chromatographia* **2018**, *81*, 1439–1444. [CrossRef]

99. Zhang, N.; Tian, M.; Liu, X.; Yang, L. Enzyme assay for d-amino acid oxidase using optically gated capillary electrophoresis-laser induced fluorescence detection. *J. Chromatogr. A* **2018**, *1548*, 83–91. [CrossRef]

100. Lee, K.J.; Mwongela, S.M.; Kottegoda, S.; Borland, L.; Nelson, A.R.; Sims, C.E.; Allbritton, N.L. Determination of Sphingosine Kinase Activity for Cellular Signaling Studies. *Anal. Chem.* **2008**, *80*, 1620–1627. [CrossRef]

101. Karkhanina, A.A.; Mecinović, J.; Musheev, M.U.; Krylova, S.M.; Petrov, A.P.; Hewitson, K.S.; Flashman, E.; Schofield, C.J.; Krylov, S.N. Direct Analysis of Enzyme-Catalyzed DNA Demethylation. *Anal. Chem.* **2009**, *81*, 5871–5875. [CrossRef] [PubMed]

102. Falkenberg, K.J.; Johnstone, R.W. Histone deacetylases and their inhibitors in cancer, neurological diseases and immune disorders. *Nat. Rev. Drug Discov.* **2014**, *13*, 673–691. [CrossRef] [PubMed]

103. Benedetti, R.; Conte, M.; Altucci, L. Targeting Histone Deacetylases in Diseases: Where Are We? *Antioxid. Redox Signal.* **2015**, *23*, 99–126. [CrossRef]

104. Zhang, Y.; Li, F.; Kang, J. Screening of histone deacetylase 1 inhibitors in natural products by capillary electrophoresis. *Anal. Methods* **2017**, *9*, 5502–5508. [CrossRef]

105. Piccard, H.; Hu, J.; Fiten, P.; Proost, P.; Martens, E.; Van den Steen, P.E.; Van Damme, J.; Opdenakker, G. "Reverse degradomics", monitoring of proteolytic trimming by multi-CE and confocal detection of fluorescent substrates and reaction products. *Electrophoresis* **2009**, *30*, 2366–2377. [CrossRef]

106. Fayad, S.; Tannoury, M.; Morin, P.; Nehmé, R. Simultaneous elastase-, hyaluronidase- and collagenase-capillary electrophoresis based assay. Application to evaluate the bioactivity of the red alga Jania rubens. *Anal. Chim. Acta* **2018**, *1020*, 134–141. [CrossRef]

107. Nehmé, R.; Atieh, C.; Fayad, S.; Claude, B.; Chartier, A.; Tannoury, M.; Elleuch, F.; Abdelkafi, S.; Pichon, C.; Morin, P. Microalgae amino acid extraction and analysis at nanomolar level using electroporation and capillary electrophoresis with laser-induced fluorescence detection. *J. Sep. Sci.* **2017**, *40*, 558–566.

108. Fayad, S.; Nehmé, R.; Lafite, P.; Morin, P. Assaying human neutrophil elastase activity by capillary zone electrophoresis combined with laser-induced fluorescence. *J. Chromatogr. A* **2015**, *1419*, 116–124. [CrossRef]

109. Chichester, K.D.; Sebastian, M.; Ammerman, J.W.; Colyer, C.L. Enzymatic assay of marine bacterial phosphatases by capillary electrophoresis with laser-induced fluorescence detection. *Electrophoresis* **2008**, *29*, 3810–3816. [CrossRef] [PubMed]
110. Guan, Y.; Zhou, G. Ultrasensitive analysis of glucose in serum by capillary electrophoresis with LIF detection in combination with signal amplification strategies and on-column enzymatic assay. *Electrophoresis* **2016**, *37*, 834–840. [CrossRef]
111. Zhou, G.; Guan, Y. An On-Column Enzyme Mediated Fluorescence-Amplification Method for Plasma Total Cholesterol Measurement by Capillary Electrophoresis with LIF Detection. *Chromatographia* **2016**, *79*, 319–325. [CrossRef]
112. Xie, W.; Xu, A.; Yeung, E.S. Determination of NAD+ and NADH in a Single Cell under Hydrogen Peroxide Stress by Capillary Electrophoresis. *Anal. Chem.* **2009**, *81*, 1280–1284. [CrossRef]
113. Pungor, E., Jr.; Hague, C.M.; Chen, G.; Lemontt, J.F.; Dvorak-Ewell, M.; Prince, W.S. Development of a functional bioassay for arylsulfatase B using the natural substrates of the enzyme. *Anal. Biochem.* **2009**, *395*, 144–150. [CrossRef] [PubMed]
114. Stege, P.W.; Messina, G.A.; Bianchi, G.; Olsina, R.A. Determination of the beta-glucosidase activity in different soils by pre capillary enzyme assay using capillary electrophoresis with laser-induced fluorescence detection. *J. Fluoresc.* **2010**, *20*, 517–523. [CrossRef] [PubMed]
115. Enayetul Babar, S.M.; Song, E.J.; Yoo, Y.S. Analysis of calcineurin activity by capillary electrophoresis with laser-induced fluorescence detection using peptide substrate. *J. Sep. Sci.* **2008**, *31*, 579–587. [CrossRef] [PubMed]
116. Zinellu, A.; Pasciu, V.; Sotgia, S.; Scanu, B.; Berlinguer, F.; Leoni, G.; Succu, S.; Cossu, I.; Passino, E.S.; Naitana, S.; et al. Capillary electrophoresis with laser-induced fluorescence detection for ATP quantification in spermatozoa and oocytes. *Anal. Bioanal. Chem.* **2010**, *398*, 2109–2116. [CrossRef] [PubMed]
117. Tabi, T.; Lohinai, Z.; Palfi, M.; Levine, M.; Szoko, E. CE-LIF determination of salivary cadaverine and lysine concentration ratio as an indicator of lysine decarboxylase enzyme activity. *Anal. Bioanal. Chem.* **2008**, *391*, 647–651. [CrossRef] [PubMed]
118. Krylova, S.M.; Koshkin, V.; Bagg, E.; Schofield, C.J.; Krylov, S.N. Mechanistic Studies on the Application of DNA Aptamers as Inhibitors of 2-Oxoglutarate-Dependent Oxygenases. *J. Med. Chem.* **2012**, *55*, 3546–3552. [CrossRef]
119. Liu, D.N.; Li, L.; Lu, W.P.; Liu, Y.Q.; Wehmeyer, K.R.; Bao, J.J. Capillary electrophoresis with laser-induced fluorescence detection as a tool for enzyme characterization and inhibitor screening. *Anal. Sci.* **2008**, *24*, 333–337. [CrossRef]
120. Li, Y.; Liu, D.; Bao, J.J. Characterization of tyrosine kinase and screening enzyme inhibitor by capillary electrophoresis with laser-induced fluoresce detector. *J. Chromatogr. B Anal. Technol. Biomed. Life Sci.* **2011**, *879*, 107–112. [CrossRef] [PubMed]
121. Woo, N.; Kim, S.K.; Kang, S.H. Multi-immunoreaction-based dual-color capillary electrophoresis for enhanced diagnostic reliability of thyroid gland disease. *J. Chromatogr. A* **2017**, *1509*, 153–162. [CrossRef]
122. Lin, L.; Liu, S.; Nie, Z.; Chen, Y.; Lei, C.; Wang, Z.; Yin, C.; Hu, H.; Huang, Y.; Yao, S. Automatic and integrated micro-enzyme assay (AImuEA) platform for highly sensitive thrombin analysis via an engineered fluorescence protein-functionalized monolithic capillary column. *Anal. Chem.* **2015**, *87*, 4552–4559. [CrossRef] [PubMed]
123. Nichols, E.R.; Craig, D.B. Single Molecule Assays Reveal Differences Between In Vitro and In Vivo Synthesized β-Galactosidase. *Protein J.* **2008**, *27*, 376–383. [CrossRef]
124. Wong, E.; Okhonin, V.; Berezovski, M.V.; Nozaki, T.; Waldmann, H.; Alexandrov, K.; Krylov, S.N. "Inject-Mix-React-Separate-and-Quantitate" (IMReSQ) Method for Screening Enzyme Inhibitors. *JACS* **2008**, *130*, 11862–11863. [CrossRef]
125. Yan, X.; Gilman, S.D. Improved peak capacity for CE separations of enzyme inhibitors with activity-based detection using magnetic bead microreactors. *Electrophoresis* **2010**, *31*, 346–352. [CrossRef]
126. Sun, X.; Gao, N.; Jin, W. Monitoring yoctomole alkaline phosphatase by capillary electrophoresis with on-capillary catalysis-electrochemical detection. *Anal. Chim. Acta* **2006**, *571*, 30–33. [CrossRef] [PubMed]
127. Bao, J.; Regnier, F.E. Ultramicro enzyme assays in a capillary electrophoretic system. *J. Chromatogr.* **1992**, *608*, 217–224. [CrossRef]

128. Nováková, S.; Van Dyck, S.; Van Schepdael, A.; Hoogmartens, J.; Glatz, Z. Electrophoretically mediated microanalysis. *J. Chromatogr. A* **2004**, *1032*, 173–184. [CrossRef]

129. Van Dyck, S.; Van Schepdael, A.; Hoogmartens, J. Michaelis-Menten analysis of bovine plasma amine oxidase by capillary electrophoresis using electrophoretically mediated microanalysis in a partially filled capillary. *Electrophoresis* **2001**, *22*, 1436–1442. [CrossRef]

130. Gautier, M.; Alain, J.P.A.; Janos, S.; William, H.; Erika, B. Neutrophil Elastase as a Target in Lung Cancer. *Anticancer Agents Med. Chem.* **2012**, *12*, 565–579.

131. Takeuchi, H.; Gomi, T.; Shishido, M.; Watanabe, H.; Suenobu, N. Neutrophil elastase contributes to extracellular matrix damage induced by chronic low-dose UV irradiation in a hairless mouse photoaging model. *J. Dermatol. Sci.* **2010**, *60*, 151–158. [CrossRef]

132. Nehmé, H.; Nehmé, R.; Lafite, P.; Routier, S.; Morin, P. Human protein kinase inhibitor screening by capillary electrophoresis using transverse diffusion of laminar flow profiles for reactant mixing. *J. Chromatogr. A* **2013**, *1314*, 298–305. [CrossRef] [PubMed]

133. Nehmé, R.; Nehmé, H.; Saurat, T.; de-Tauzia, M.-L.; Buron, F.; Lafite, P.; Verrelle, P.; Chautard, E.; Morin, P.; Routier, S.; et al. New in-capillary electrophoretic kinase assays to evaluate inhibitors of the PI3k/Akt/mTOR signaling pathway. *Anal. Bioanal. Chem.* **2014**, *406*, 3743–3754. [CrossRef] [PubMed]

134. Řemínek, R.; Zeisbergerová, M.; Langmajerová, M.; Glatz, Z. New capillary electrophoretic method for on-line screenings of drug metabolism mediated by cytochrome P450 enzymes. *Electrophoresis* **2013**, *34*, 2705–2711. [CrossRef]

135. Arrieta, J.M.; Herndl, G.J. Assessing the Diversity of Marine Bacterial β-Glucosidases by Capillary Electrophoresis Zymography. *Appl. Environ. Microbiol.* **2001**, *67*, 4896–4900. [CrossRef]

136. Wild, S.; Roglic, G.; Green, A.; Sicree, R.; King, H. Global Prevalence of Diabetes: Estimates for the year 2000 and projections for 2030. *Diabetes Care* **2004**, *27*, 1047–1053.

137. Danaei, G.; Finucane, M.M.; Lu, Y.; Singh, G.M.; Cowan, M.J.; Paciorek, C.J.; Lin, J.K.; Farzadfar, F.; Khang, Y.-H.; Stevens, G.A.; et al. National, regional, and global trends in fasting plasma glucose and diabetes prevalence since 1980: Systematic analysis of health examination surveys and epidemiological studies with 370 country-years and 2·7 million participants. *Lancet* **2011**, *378*, 31–40. [CrossRef]

138. Lowry, O.H.; Passonneau, J.V.; Schulz, D.W.; Rock, M.K. The measurement of pyridine nucleotides by enzymatic cycling. *J. Biol. Chem.* **1961**, *236*, 2746–2755. [PubMed]

139. Ying, W.; Garnier, P.; Swanson, R.A. NAD+ repletion prevents PARP-1-induced glycolytic blockade and cell death in cultured mouse astrocytes. *Biochem. Biophys. Res. Commun.* **2003**, *308*, 809–813. [CrossRef]

140. Guetschow, E.D.; Steyer, D.J.; Kennedy, R.T. Subsecond Electrophoretic Separations from Droplet Samples for Screening of Enzyme Modulators. *Anal. Chem.* **2014**, *86*, 10373–10379. [CrossRef]

141. Gong, M.; Kim, B.Y.; Flachsbart, B.R.; Shannon, M.A.; Bohn, P.W.; Sweedler, J.V. An On-Chip Fluorogenic Enzyme Assay Using a Multilayer Microchip Interconnected with a Nanocapillary Array Membrane. *IEEE Sens. J.* **2008**, *8*, 601–607. [CrossRef]

142. Ohla, S.; Beyreiss, R.; Scriba, G.K.E.; Fan, Y.; Belder, D. An integrated on-chip sirtuin assay. *Electrophoresis* **2010**, *31*, 3263–3267. [CrossRef]

143. Belder, D.; Ludwig, M.; Wang, L.W.; Reetz, M.T. Enantioselective catalysis and analysis on a chip. *Angew. Chem. (Int. Ed. Engl.)* **2006**, *45*, 2463–2466. [CrossRef]

144. Ma, J.; Zhang, L.; Liang, Z.; Shan, Y.; Zhang, Y. Immobilized enzyme reactors in proteomics. *TrAC Trends Anal. Chem.* **2011**, *30*, 691–702. [CrossRef]

145. Asanomi, Y.; Yamaguchi, H.; Miyazaki, M.; Maeda, H. Enzyme-immobilized microfluidic process reactors. *Molecules* **2011**, *16*, 6041–6059. [CrossRef] [PubMed]

146. Qiao, J.; Qi, L.; Ma, H.; Chen, Y.; Wang, M.; Wang, D. Study on amino amides and enzyme kinetics of L-asparaginase by MCE. *Electrophoresis* **2010**, *31*, 1565–1571. [CrossRef]

147. Qiao, J.; Qi, L.; Mu, X.; Chen, Y. Monolith and coating enzymatic microreactors of l-asparaginase: Kinetics study by MCE–LIF for potential application in acute lymphoblastic leukemia (ALL) treatment. *Analyst* **2011**, *136*, 2077–2083. [CrossRef]

148. Chen, H.-X.; Busnel, J.-M.; Gassner, A.-L.; Peltre, G.; Zhang, X.-X.; Girault, H.H. Capillary electrophoresis immunoassay using magnetic beads. *Electrophoresis* **2008**, *29*, 3414–3421. [CrossRef]

149. Wang, H.; Dou, P.; Lü, C.; Liu, Z. Immuno-magnetic beads-based extraction-capillary zone electrophoresis-deep UV laser-induced fluorescence analysis of erythropoietin. *J. Chromatogr. A* **2012**, *1246*, 48–54. [CrossRef]

150. Morales-Cid, G.; Diez-Masa, J.C.; de Frutos, M. On-line immunoaffinity capillary electrophoresis based on magnetic beads for the determination of alpha-1 acid glycoprotein isoforms profile to facilitate its use as biomarker. *Anal. Chim. Acta* **2013**, *773*, 89–96. [CrossRef]

151. Donghui, Y.; Pierre Van, A.; Stephanie, P.; Bertrand, B.; Jean-Michel, K. Enzyme Immobilized Magnetic Nanoparticles for In-Line Capillary Electrophoresis and Drug Biotransformation Studies: Application to Paracetamol. *Comb. Chem. High Throughput Screen* **2010**, *13*, 455–460.

152. Stege, P.W.; Raba, J.; Messina, G.A. Online immunoaffinity assay-CE using magnetic nanobeads for the determination of anti-Helicobacter pylori IgG in human serum. *Electrophoresis* **2010**, *31*, 3475–3481. [CrossRef] [PubMed]

molecules

MDPI

Review

Advances in the Biology of Seed and Vegetative Storage Proteins Based on Two-Dimensional Electrophoresis Coupled to Mass Spectrometry

Daniel Mouzo [1], Javier Bernal [1], María López-Pedrouso [1], Daniel Franco [2] and Carlos Zapata [1,*]

[1] Department of Zoology, Genetics and Physical Anthropology, University of Santiago de Compostela, 15782 Santiago de Compostela, Spain; daniel.mouzo.calzadilla@usc.es (D.M.); javier.bernal.pampin@gmail.com (J.B.); mariadolores.lopez@usc.es (M.L.-P.)
[2] Meat Technology Center of Galicia, 32900 San Cibrao das Viñas, Ourense, Spain; danielfranco@ceteca.net
* Correspondence: c.zapata@usc.es; Tel.: +34-8-8181-6922

Academic Editors: Angela R. Piergiovanni and José Manuel Herrero-Martínez
Received: 31 July 2018; Accepted: 21 September 2018; Published: 26 September 2018

Abstract: Seed storage proteins play a fundamental role in plant reproduction and human nutrition. They accumulate during seed development as reserve material for germination and seedling growth and are a major source of dietary protein for human consumption. Storage proteins encompass multiple isoforms encoded by multi-gene families that undergo abundant glycosylations and phosphorylations. Two-dimensional electrophoresis (2-DE) is a proteomic tool especially suitable for the characterization of storage proteins because of their peculiar characteristics. In particular, storage proteins are soluble multimeric proteins highly represented in the seed proteome that contain polypeptides of molecular mass between 10 and 130 kDa. In addition, high-resolution profiles can be achieved by applying targeted 2-DE protocols. 2-DE coupled with mass spectrometry (MS) has traditionally been the methodology of choice in numerous studies on the biology of storage proteins in a wide diversity of plants. 2-DE-based reference maps have decisively contributed to the current state of our knowledge about storage proteins in multiple key aspects, including identification of isoforms and quantification of their relative abundance, identification of phosphorylated isoforms and assessment of their phosphorylation status, and dynamic changes of isoforms during seed development and germination both qualitatively and quantitatively. These advances have translated into relevant information about meaningful traits in seed breeding such as protein quality, longevity, gluten and allergen content, stress response and antifungal, antibacterial, and insect susceptibility. This review addresses progress on the biology of storage proteins and application areas in seed breeding using 2-DE-based maps.

Keywords: seed proteomics; seed phosphoproteomics; seed glycoproteomics; seed quality traits; seed molecular breeding

---

1. Introduction

Storage proteins accumulate during seed development within membrane-bound organelles called protein bodies and serve as a reservoir of amino acids, reduced nitrogen, carbon, and sulfur required for germinating seedlings [1–5]. Storage proteins also play a crucial role in human nutrition and livestock feed. Plants provide most (ca. 58%) of the dietary protein consumed worldwide compared to animal-based protein sources, although with marked variations depending on the region and economic status [6–9]. In particular, seeds are a major source of the dietary protein content that varies approximately from 10% (dry weight) in cereals to 40% in some legumes and oilseeds [1]. Storage proteins determine to a great extent the seed nutritional quality because they account for a major part of the total protein content. By way of illustration, approximately 70–80% of the total amount of reduced nitrogen in cereals and legume grains can be attributed to seed storage proteins (SSPs) [10]. In addition, some SSPs and vegetative storage proteins (VSPs) can exhibit additional enzymatic activities such as lipid acyl hydrolase, acyltranferase, esterase and acid phosphatase activities capable of assuming useful supplementary biological functions, including defense and antioxidant functions [11–14].

The model species *Arabidopsis thaliana* L. has played a key role in identifying gene regulatory networks that govern seed development and germination. A wide repertoire of genetic technologies enabled the identification of essential regulatory genes during seed development and germination in Arabidopsis as well as the identification of orthologous genes in other plant species [15–20]. These technologies include forward genetic screens of lines obtained by T-DNA insertional mutagenesis for tagged mutants that produce a knockout phenotype, microarray RNA transcriptional profiling, and identification of seed-specific transcription factors (TFs). Genes involved in the regulatory networks responsible for the synthesis, accumulation and mobilization of seed storage proteins have been identified in Arabidopsis and other plants [20,21]. Dormancy induction and germination are greatly regulated by the dynamic balance between the functional antagonist abscisic acid (ABA) and gibberelic acid (GA) phytohormones [22]. Considerable progress has been achieved in unraveling the regulatory mechanisms underlying ABA response [23–26]. In particular, a number of protein-coding genes and TFs have been associated with the hormonal regulation involved in the synthesis and accumulation of storage proteins [20].

Seed proteome comprises a heterogeneous collection of functionally differentiated proteins that undergo highly dynamic qualitative and quantitative changes in order to meet seed requirements during development and germination. Storage proteins are typically multimeric proteins encoded by multi-gene families constituted by highly homologous genes clustered on one or various chromosomes [14,20,27,28]. They often undergo abundant glycosylations and phosphorylations, two types of co- and/or post-translational modifications (PTMs) that notably increase the diversity of isoforms [29,30]. Proteomics encompasses a wide range of technologies with sufficient potential for the detailed characterization of the broad set of storage protein isoforms. There have been a large number of gel-based and gel-free MS-driven proteomic studies focused on seed proteome [31–37]. The 2-DE proteomic technology initially developed by O'Farrel [38] opened the way to numerous studies addressing the characterization of storage proteins. Reference maps of many storage proteins have been constructed based on the separation of total seed proteins by 2-DE and protein identification by downstream MS analysis.

2-DE-based maps of storage proteins have been obtained using two different experimental strategies with strengths and weaknesses. Hundreds of publications have used experimental protocols for the study of global seed proteins with very different relative abundance [31–35,39–41]. This is an optimal experimental approach to assess the interplay between storage proteins and other seed proteins, but it entails the loss of definition of storage protein isoforms on 2-DE gels. Alternatively, a minority of studies used 2-DE specific protocols aimed at obtaining high-resolution profiles of storage proteins [29,30,42–44]. This approach is very useful to characterize storage protein isoforms and their response to internal and external seed stimuli at higher level of resolution, although the information it provides is decoupled from the rest of seed proteins. Overall, the application of these two strategies has provided most of the advances in the biology of storage proteins. These advances cover facets as diverse as the identification of isoforms and their relative abundance, the identification, mapping and quantitation of phosphorylated and glycosylated isoforms and the assessment of qualitative and quantitative changes of isoforms during seed development and germination. Seed breeding programs have benefited from these advances for the improvement of many seed traits of interest such as protein quality, longevity, gluten and allergen content, stress response and antifungal, antibacterial and insect susceptibility [45–51].

This review focuses on the use and importance of 2-DE-based maps to obtain insights into the biology of storage proteins and application areas in seed breeding.

2. Terminology and Classification of Storage Proteins

SSPs are currently denominated according to profoundly heterogeneous criteria: extraction/solubility in distinct solvents (e.g., albumins), sedimentation coefficients (e.g., 7S), generic names in Latin (e.g., hordeins from barley, *Hordeum vulgare* L.), trivial names (cactin from *Cereus jamacaru* DC.) and specific terminology for polypeptide subunits encoded by multigene families (e.g., phaseolin α-type polypeptide from common bean, *Phaseolus vulgaris* L.) [34]. However, most storage proteins have traditionally been classified into four main groups on the basis of their solubility in different solvents as proposed by Osborne [52]: water (albumins), dilute saline (globulins), alcohol-water mixtures (prolamines) and dilute acid or alkali (glutelins). New bioinformatics algorithms have recently been proposed for a higher classification accuracy using specific sequences available in public databases [53,54].

VSPs are a differentiated set of plant storage proteins located in vegetative tissues (tubers, stems, roots or leaves) of plants such as the sweet potato (sporamins), the potato (patatins) and *Oxalis tuberosa* Mol. (ocatins) [2,13,55,56]. For example, the patatin multigene family can be divided into class-I and class-II gene subfamilies with differential tissue expression patterns: class I transcripts are potato (*Solanum tuberosum* L.) tuber specific while class II transcripts are expressed not only in tubers but also in roots but much less abundant than class I transcripts [57,58]. VSPs are not grouped together with SSPs because they belong to a family of unrelated proteins and exhibit certain different characteristics such as a distinct form of mobilization [2,13,55,59].

A representative list of storage proteins (SSPs and VSPs) that includes important worldwide agricultural crops is shown in Table 1.

Table 1. List of seed and vegetative storage proteins in different crop types.

Crop	Storage Proteins		Percentage of Total Protein	Molecular Weight Subunits (kDa)	References
Maize (*Zea mays* L.)	Globulins		12–16		[3,60–64]
		globulin-1		63, 45, 26, 23	
		globulin-2			
	Prolamins		50–70		
		α-zeins	25–49	22, 19	
		β-zeins	1–4	14–16	
		γ-zeins	6–13	27, 16, 50	
		δ-zeins	1–4	10	
Wheat (*Triticum aestivum* L.)	Prolamins		80		[65–73]
	gliadins		30–50	30–80	
		α-gliadins	15–30		
		β-gliadins			
		γ-gliadins			
		ω-gliadins			
	glutenins				
		LMW-GS	12	42–51 (B), 30–40 (C), 58 (D)	
		HMW-GS		80–130 (A)	
	Globulins				
		11-12S triticins	5	58 (D), 22 (δ), 52 (A), 23 (α)	
Rice (*Oryza sativa* L.)	Glutelins		60–80	35–40, 20–22	[74–76]
	Prolamins		20–30	10, 13, 16	
	Globulins				
		α-globulins	2–8	26	

Table 1. *Cont.*

Crop	Storage Proteins	Percentage of Total Protein	Molecular Weight Subunits (kDa)	References
Potato (*Solanum tuberosum* L.)	Patatins	45	39–45	[30,77,78]
	Kunitz protease inhibitors		20	
	Protease inhibitors 1		45	
	Protease inhibitors 2			
	Carboxypeptidase inhibitors		10	
	Lipoxygenases		97	
Soybean (*Glycine max* L.)	Globulins			[79–81]
	α-conglycinins			
	7S vicilin/ β-conglycinins γ-conglycinins	40	76 (α), 72 (α′), 52 (β)	
	11S legumin/ glycinins	25	56 (G1), 54 (G2), 54 (G3), 64 (G4), 58 (G5)	
Barley (*Hordeum vulgare* L.)	Prolamins			[68,82]
	hordeins	35–55		
	B-hordeins	15–44	30–45	
	C-hordeins	4–11	45–75	
	D-hordeins		45	
	γ-hordeins			
Sunflower (*Helianthus annuus* L.)	Globulins		37–43 (α), 31–35 (α′), 21–30 (β)	[83–85]
	11S helianthinins	38		
	Albumins			
	2S	62	12–20	
Common Bean (*Phaseolus vulgaris* L.)	Globulins			[14,44,86,87]
	7S phaseolins	40–50		
	11S legumins	3		
	Lectins			
	phytohemagglutinins	5–10		
	α-amylase inhibitors			
Oat (*Avena sativa* L.)	Globulins	10–55		[71,88,89]
	3S		48–52	
	7S		50–70	
	11S		60	
	12S avenalins		32–43 (α), 19–25 (β)	
	Albumins	10–20		
	Prolamins	12–14		
	Glutelins	23–54		
Pea (*Pisum sativum* L.)	Globulins			[90,91]
	7S vicilins		47, 50, 34, 30	
	11S legumins		41 (α), 22 (β), 23 (β′)	
	convincilins		78, 72	
Chickpea (*Cicer arietinum* L.)	Albumins			[89,92]
	2S	12		
	Globulins	50		
	7S vicilins			
	11S legumins		40–47 (α), 24–25 (β)	
	Glutelins	18.1		
	Prolamins	2.8		
Pomegranate (*Punica granatum* L.)	Globulins	40.5	38–54, 13–18	[93]
	Albumins	32.2	58–116, 33–46, 15–23	
	Glutelins	15.6	37, 21–23, 14	
	Prolamins	9.7	15, 20, 24	

Table 1. *Cont.*

Crop	Storage Proteins		Percentage of Total Protein	Molecular Weight Subunits (kDa)	References
Lentils (*Lens culinaris* Medik.)	Globulins				[94]
		11S legumins	21	38–43	
		7S vicilin/ convicilins	72	15–59	
	Albumins				
		2S			
Rapeseed (*Brassica napus* L)	Globulins				[95]
		12S cruciferins	20	29–33 (α), 21–23 (β)	
	Albumins				
		2S napins	60	4–9	

Globulins predominate in dicotyledoneous seeds whereas prolamins are the major storage proteins in most cereals. Globulins are located in the embryo and outer aleurone layer of the endosperm and are commonly divided according to their different sedimentation coefficients (7S and 11S). They are very similar to 7S vicilins in legumes and other dicotyledoneous plants [3]. In maize (*Zea mays* L.), globulins are classified as globulin-1, the most abundant storage protein in embryos, and globulin-2. In soybean (*Glycine max* L.), the seeds contain a considerable amount of globulins, namely β-conglycin (7S globulin) and glycinin (11S globulin). β-conglycin has a trimeric structure composed of α, α', and β subunits with molecular weights ranging from 50 to 76 kDa. Glycinins consist of six subunits linked by disulfide bonds, but the five major subunits are G1–G5 whose molecular weights range from 54 to 64 kDa. Prolamins are the major proteins in endosperm and they are more variable than globulins. In maize grain, zeins are the most abundant storage proteins and are mainly accumulated in the endosperm between 12 and 40 days after pollination [96]. They are grouped into α (19 and 22 kDa), β (16 kDa), γ (16, 27 and 50 kDa) and δ (10 kDa) zeins [63]. Wheat (*Triticum aestivum* L.) prolamins, gliadins and glutenins form gluten and are located in grain endosperm. Gliadins are often subdivided into various subtypes in accordance with their electrophoretic mobilities (i.e., α-, β-, γ- and ω-gliadins), whereas glutenin subunits are subdivided according to their molecular weights (i.e., HMW-GS and LMW-GS glutenins). In rice (*Oryza sativa* L.), glutelins are the major seed storage proteins that contain hexamers of α-polypeptides (35 kDa) and β-polypeptides (22 kDa). Storage proteins are abundant proteins but subtypes are differentially represented in seed/tuber proteomes with relative amounts ranging from 1–4% (δ-zeins, *Z. mays*) to 72% (7S lentil vicilins, *Lens culinaris* Medik.), whereas M_r-values range from 10 (δ-zeins, *Z. mays*) to 130 kDa (HMW-GS glutenins, *T. aestivum*) (Table 1).

3. Two-Dimensional-Based Reference Maps of Storage Proteins

2-DE can be routinely applied for the separation of highly complex mixtures of proteins from cell, tissue, organ and organism protein extracts in accordance with their isoelectric point (*pI*) and molecular mass (M_r) in two successive steps: isoelectric focusing (IEF) in the first dimension and sodium dodecyl sulphate-polyacrylamide gel electrophoresis (SDS-PAGE) to resolve denatured proteins in the second. The introduction of immobilized pH gradients (IPGs) using bifunctional immobiline reagents enabled us to obtain highly stable pH gradients in the first dimension increasing resolution, reproducibility, the detection of lower abundance proteins and the separation of highly acidic and alkaline proteins [97,98]. Many other technical achievements contributed to the optimization of 2-DE, such as more efficient protein extraction methods, the running of multiple gels in parallel, highly sensitive protein stain methods based on fluorescent dyes compatible with subsequent protein identification by MS technologies, and advanced computer software for the analysis of gel images [99–103]. Technical inter-gel variation of protein spots can be reduced using an internal pooled standard in multiplexing methods. Difference gel electrophoresis (DIGE) enables the simultaneous running of up to three different samples in a single 2-DE gel using pre-electrophoretic labeling of protein samples with distinct spectrally-resolvable fluorescent CyDyes [101].

Dedicated protein extraction protocols can alleviate in part some of the limitations of the standard 2-DE system, including the analysis of low-abundant proteins and membrane proteins [98,102,104,105]. It is noteworthy that plant tissues contain relatively lower amounts of proteins than other organisms and a large number of biological compounds that interfere notably with the extraction, solubilization and separation of proteins by 2-DE, such as cell walls, lipids, polysaccharides, polyphenols and large quantities of proteases. Therefore, protein extraction is the initial and one of the most critical steps in plant proteomic studies because it determines to a large extent the final quality of 2-DE [99,106–109]. Overall, 2-DE is a laborious and poorly automated technology that requires a great deal of expertise to successfully exploit its potential.

High-resolution 2-DE can successfully separate, detect and quantify up to thousands of proteins simultaneously [99]. It is routinely applied in current proteomics to effectively analyze abundant and soluble proteins with an amount of 1–2 ng per spot expressed at greater than 10^3 copies/cell, a linear dynamic range about three orders of magnitude, molecular mass ranging from 15 to 150 kDa and pH intervals from 2.5 to 12 [99,104,110]. Accordingly, 2-DE has enough resolving power to separate most of the isoforms of storage proteins. These proteins are soluble and highly abundant, exhibiting a range of M_r and pI within 2-DE resolution limits. By way of illustration, values of M_r over phaseolin and patatin isoforms have a range of variation between 40 and 50 kDa, whereas pI-values range from 4.5 to 5.8 [29,30,42,44]. Gel location of storage protein isoforms can be initially established in accordance with their theoretical M_r and pI values and candidate protein spots eventually confirmed by MS for polypeptide/protein identification. 2-DE has the important ability to detect degraded proteins by comparing their M_r values observed on gels to those corresponding theoretical values [109,111].

High-resolution profiles for storage proteins can be achieved by conveniently adjusting the amount of total protein loaded onto IPG strips [29,30,42–44]. Figure 1 shows standard and optimized phaseolin and patatin profiles by loading low amounts of total protein extracts from common bean seeds and potato tubers, respectively. It can be seen that dedicated 2-DE protocols produce good quality gel images with well-focused and separate protein spots corresponding to different phaseolin and patatin isoforms. 2-DE phaseolin and patatin profiles comprise a large number of spots organized in a compact way on the same gel region. Protein storage profiles can also exhibit multiple constellations of spots widely distributed on 2-DE gels (Figure 2). Dedicated 2-DE protocols have the additional advantage that the statistical cost by probability adjustments for multiple hypothesis testing is lower than in protocols addressed to the analysis of total seed proteomes, which leads to an increase in the statistical power of significance tests.

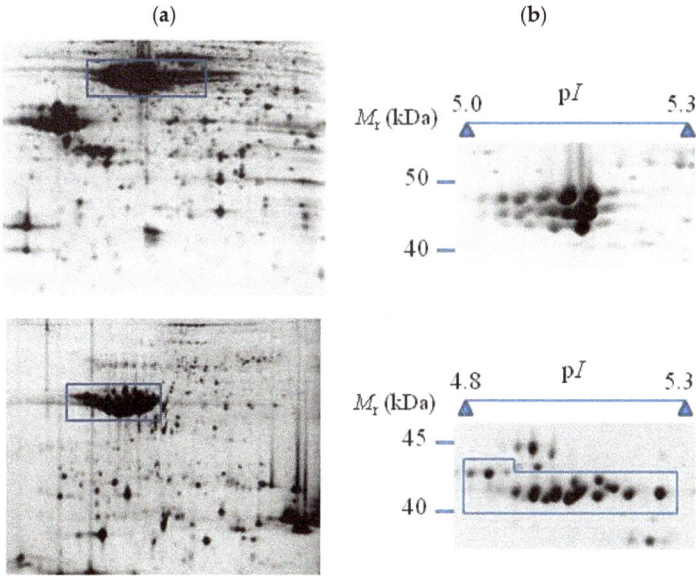

Figure 1. Standard (**a**) and targeted (**b**) 2-DE gel images of phaseolin (above) and patatin (below) isoforms from common bean (*P. vulgaris*) seeds and potato (*S. tuberosum*) tubers. Standard 2-DE gels were obtained from 250 µg of total seed protein or total tuber protein extracts loaded into 24-cm-long IPG strips of linear pH gradient 4–7 in the first dimension. The second dimension (SDS-PAGE) was run on 12% (*w/v*) SDS-PAGE gels. Gels were subsequently stained with SYPRO Ruby fluorescent stain. Targeted 2-DE gel images for high-resolution profiles were obtained under the same conditions but using only 75 µg of total protein extracts.

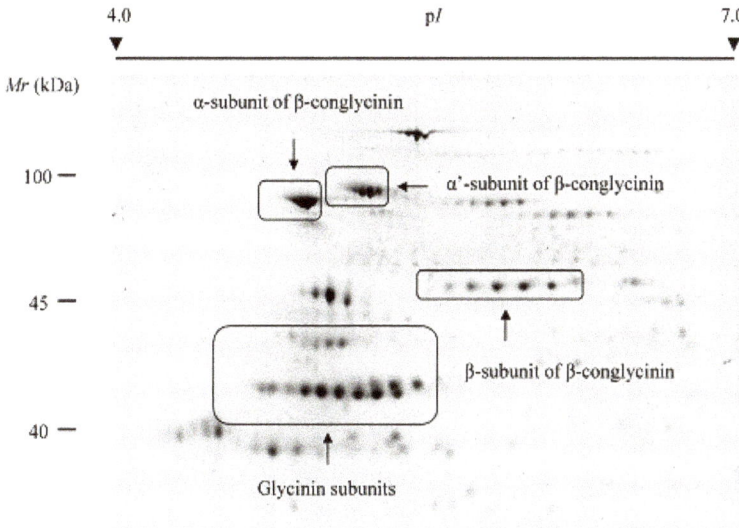

Figure 2. Gel image of high-resolution profile of soybean (*G. max*) storage proteins (β-conglycinin and glycinin subunits) obtained by the targeted 2-DE.

2-DE is particularly useful for identification of PTMs that change the pI and/or M_r of proteins such as phosphorylations and glycosylations [102,112]. 2-DE-based reference maps of storage proteins can, therefore, be implemented with in-gel detection and mapping of phosphorylated and glycosylated isoforms (Figure 3). The Pro-Q diamond phosphoprotein stain (Pro-Q DPS) is a simple, direct, rapid and commonly used method for in-gel multiplex detection, mapping and quantitation of phosphorylated proteins [113,114]. Recent studies indicate, however, that the phosphoprotein chemical dephosphorylation of seed protein extracts with hydrogen fluoride-pyridine (HF-P) [115] prior to 2-DE is a highly valuable strategy for more accurate in-gel quantitation of phosphorylated storage proteins [29,30]. Phosphorylation levels for 2-DE spots can be directly assessed from volume changes between dephosphorylated and control sample profiles.

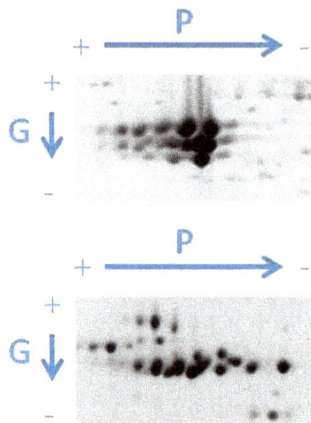

Figure 3. Gel images of differentially phosphorylated (P) and glycosylated (G) isoforms of phaseolin (above) and patatin (below) obtained by targeted 2-DE.

The analysis of phosphorylated isoforms of storage proteins based on dedicated 2-DE maps has several major advantages in comparison to MS-driven analyses. A standard "bottom up" quantitative phosphoproteomics workflow involves the enzymatic or chemical digestion of a mixture of proteins into peptides to produce MS/MS spectra [33]. The redundancy of peptides and phosphorylation sites over high sequence identity protein isoforms hinders the assignation of specific peptides to a single isoform [112]. It is noteworthy that storage protein isoforms are encoded by gene families that exhibit high sequence identity mainly due to concerted evolution mechanisms of unequal crossing over and gene conversion [116]. For instance, patatin isoforms are encoded by a multigene family constituted by ~10–18 genes per haploid genome [28] and exhibit a sequence homology of at least 90% [58,117]. In addition, many other factors can lead to erroneous conclusions in MS-driven PTM analysis such as the co-elution of peptides, the loss of phosphoryl group during ionization process and phosphate transfer to acceptor residues, a lack of reproducibility and a low number of commonly used biological replicates [96,112,118]. It is noteworthy that some of these methodological constraints apply to powerful MS-based methods used for quantitative proteomics such as stable isotope labeling with amino acids in cell culture (SILAC) and isobaric tags for relative and absolute quantitation (iTRAQ). On the other hand, phosphopeptide enrichment strategies are usually accomplished prior to MS analysis because of the fact that many phosphoproteins/phosphopeptides from biological samples may be present in substoichiometric amounts [33,118]. In the case of storage proteins, the application of enrichment methods is not required because they are abundantly phosphorylated proteins [29,30]. Phosphopeptide enrichment methods such as immobilized metal affinity chromatography (IMAC) and titanium dioxide (TiO$_2$) impair the evaluation of quantitative changes in the phosphorylation status among storage protein isoforms, although they are very useful for phosphosite identification.

Finally, the detection and quantitation of glycosylated isoforms of storage protein can be assessed by different methods, including the enzymatic deglycosylation of total protein extracts [42], in-gel glycoprotein-specific Pro-Q Emerald fluorescent stain [119] and glycopeptide enrichment using a zwitterionic (ZIC) hydrophilic interaction liquid chromatography (HILIC) column or affinity chromatography on a concanavalin-A-sepharose column [120,121]. Storage protein glycoforms can be identified efficiently by their M_r shifts on gels using targeted 2-DE protocols [30,42]. Glycosylated peptides are often difficult to identify in MS analyses because glycosylations change the hydrophobicity/hydrophilicity of the peptide [110].

4. Advances in the Biology of Storage Proteins

An exhaustive number of studies using 2-DE-based maps have contributed significantly to the characterization of the wide diversity of types, subunits and isoforms of storage proteins, their relative abundance in seeds and tubers, PTMs, targeted mutation effects and both qualitative and quantitative variations within and between wild and cultivated accessions [29–35,39–44]. In addition, 2-DE-based maps have provided valuable information on the complex dynamic changes of storage proteins during seed development and germination.

4.1. Seed Development

The available evidence indicates that storage proteins accumulate following variable patterns during embryo growth and seed filling, depending on the type of storage protein and cultivar. Thus, Gallardo et al. [4] reported that the major storage proteins 11S legumins and 7S vicilins of the model legume *Medicago truncatula* L. are synthesized in a specific temporal order and accumulated in different relative amounts during seed development. Analysis of protein abundance changes during time course were assessed by 2-DE and protein identification by MALDI-TOF and nano-LC-MS/MS sequencing. Interestingly enough, they also found a parallel evolution in the expression of the *pII* gene involved in the regulation of the synthesis of the amino acid arginine needed for storage protein synthesis using a transcriptomics dataset. Guo et al. [67] reported that five types of wheat storage proteins (i.e., γ-gliadins, globulins, avenin-like proteins, triticins and LMW-S glutenin subunits) accumulated differentially during grain development using 2-DE and tandem MALDI-TOF/TOF MS. This study also showed that LMW-S glutelin subunits and triticins exhibited differential abundance in two Chinese bread wheat cultivars at late seed development stages. In contrast, storage proteins of rapeseed (*Brassica napus* L.), i.e., napins, cruciferins and oleosins, were found to be accumulated only during the early and middle stages of seed growth by applying histochemical and inmunostaining techniques [20,122]. In addition, recent studies have revealed that different isoforms of phaseolin/patatin are differentially accumulated during seed/tuber development within and among cultivars from quantitative analysis of phaseolin/patatin isoforms using dedicated 2-DE protocols and protein identification by MALDI-TOF/TOF MS [29,30,42–44]. Taken together, these observations raise the question of the molecular and biochemical mechanisms responsible for differential accumulation of storage protein isoforms during seed/tuber development, but they also suggest that the differential accumulation of storage proteins and isoforms has a significant meaning for their mobilization in the germination stage.

Reversible phosphorylation is the most ubiquitous and well-studied type of PTM that regulates a huge variety of key biological processes, including cell cycle, metabolism, subcellular locatization, apoptosis, and signal transduction pathways [33,123,124]. The analysis of temporal phosphorylation changes in storage proteins is of paramount importance to unraveling their functional role at different stages of seed development. In recent decades the number of reports on the phosphorylation of different SSPs during seed development, dormancy and germination has greatly increased: globulins (7S- and 11S-globulins, 12S cruciferin, 12S triticin, cupin and globulin 3) in Arabidoposis, common bean, rapeseed, rice, Scots pine (*Pinus sylvestris* L.), sunflower (*Helianthus annuus* L.) and wheat [29,67,83,125–132]; prolamins in wheat [133,134]; albumins (2S napin) in Scots pine and

Arabidopsis [126,127,130]; and glutelins in rice [135]. Bernal et al. [30] have recently reported the first evidence for the phosphorylation of VSPs in patatin.

Identification and profiling studies of phosphorylated storage proteins based on 2-DE maps combined with various other techniques are listed in Table 2. Phosphoproteomic studies show that storage proteins are abundantly phosphorylated and may play a key role during seed development. Meyer et al. [129] reported a large-scale MS-based study of enriched subproteome of phosphoproteins by the IMAC method at five sequential stages (2–6 weeks after flowering) of seed development in soybean, rapeseed and Arabidopsis. A total of 2001 phosphopeptides and 1026 unambiguous phosphorylation sites were identified across 956 non-redundant proteins, including storage proteins. Interestingly, a considerable fraction (25%) of phosphoproteins consisted of storage proteins that contained the X-S-D-X phosphorylation motif. Targeted 2-DE-based maps coupled to the chemical method of dephosphorylation with HF-P have shown high phosphorylation levels in storage protein isoforms. Phosphorylation rates over phaseolin isoforms in dormant common bean seeds (two cultivars) and patatin isoforms from mature potato tubers (one cultivar) measured by the *PR* coefficient averaged 46–63% and 34%, respectively [29,30]. Furthermore, in silico phosphopeptide analysis also revealed the occurrence of a putative phosphosite in phaseolin phosphopeptides encompassing sequence X-S-D-X in the phaseolin. This peptide, therefore, appears to be a general target for phosphorylation during seed development.

2-DE-based maps show that the accumulation of phosphorylated storage protein isoforms during seed filling also follows variable patterns. Agrawal and Thelen [125] performed the first comprehensive study aimed at detecting and quantifying phosphoproteins in development seeds. More specifically, phosphoprotein profiling was performed in rapeseed through the same five sequential phases of seed development as Meyer et al. [129] by means of 2-DE-based maps coupled to in-gel phosphoprotein specific staining with Pro-Q DPS fluorescent dye and LC-MS/MS for protein and phosphorylation site identification. The results of the study showed that 40% of phosphorylated cruciferin subunits increased during seed filling process, whereas the remaining phosphorylated subunits generally decreased with seed development. Meyer et al. [129] also reported that some phosphorylated cruciferin subunits were over-represented in the late maturation stage of seed development. Dedicated 2-DE protocols have disclosed that phosphorylation rates (PR) across different phaseolin/patatin isoforms from dormant seed/tuber were in the range of 13–82% and 5–52%, respectively [29,30].

The complex regulatory mechanisms underlying dynamic changes in the phosphorylation status of storage proteins in response to seed development and environmental factors are not yet sufficiently known. However, it is assumed that the interplay of protein kinases, protein phosphatases and phytohormones participates in the signaling and metabolic networks that control the phosphorylation/dephosphorylation levels of storage proteins. The CK2 protein is a Ser/Thr kinase presents in all eukaryotes and has pleiotropic effects; it is also involved in the regulation of multiple plant growth and development processes and ABA signalling [136–138]. Irar et al. [126] used 2-DE-based maps for the phosphoproteome profiling of heat-stable proteins from Arabidopsis dry seeds and phosphoaffinity chromatography for phosphoprotein enrichment. They reported several probable hits of phosphorylation in storage and like-storage proteins, and an increased probability of phosphorylation of serine over threonine residues by CK2, using in silico prediction of phosphorylation sites from MALDI-TOF MS and LC MS/MS data. On the other hand, the ABA-insensitive 1 (ABI1) protein phosphatase is a negative regulator of the ABA signal and interacts with proteins linked to the ubiquitin-proteosome system (UPS) [139,140]. Wan et al. [127] showed that cruciferins of *A. thaliana* may be an in vivo target for ABI1 during seed development and provided evidence that cruciferin phosphorylation levels might be regulated by ABI1 using 2-DE maps coupled to immunological detection against phosphorylated cruciferin. They also found that cruciferins had differential levels of Tyr phosphorylation in mutant *ABI1* and wild types, which suggests that Tyr phosphorylation is involved in ABA signaling.

4.2. Seed Germination

2-DE-based proteomic analyses revealed that the accumulation of storage proteins can still proceed in late stages of seed development and the onset of germination. Chibani et al. [141] reported that cruciferin precursors in Arabidopsis are accumulated by de novo synthesis during late stages of seed development leading to dormancy breakage. The accumulation of cruciferin precursors was documented by 2-DE following protein identification by MALDI-TOF MS. Proteomic research on Arabidopsis seed dormancy by 2-DE coupled to MALDI-TOF MS from seeds of the GA-deficient ga1 mutant and wild-type seeds treated with a specific inhibitor of GA biosynthesis suggests that GA is involved in the processing of precursor forms of storage proteins and accumulation of processed forms in mature seeds [142]. The comparison of 2-DE patatin profiles in dormant tubers and the onset of germination led to a better understanding of the metabolic status of storage proteins after the dormancy break. Lehesranta et al. [143] reported temporal differences of patatin abundance throughout the potato tuber lifecycle (cv. Desirée). More specifically, it was found that most patatin isoforms increase during development, are present in high amounts at the onset of sprouting (i.e., sprouts ca. 1 cm long) and remain approximately constant until tubers are fully sprouted (i.e., sprouts ca. 20 cm long) when patatin abundance decreases. Accordingly, analyses on transcripts encoding patatin throughout the potato tuber cycle based on cDNA-AFLP fingerprinting and expressed sequence tag (EST) libraries have shown that patatin transcripts are still expressed at the onset of tuber sprouting [144,145]. Similar results have been reported after chemically (bromoethane) induced cessation of dormancy using microarrays constructed from potato EST libraries [146]. Overall, these studies suggest that the major tuber storage protein encoded by the patatin multigene family is also synthesized after the dormancy break to ensure growth of the developing sprout.

Changes in the abundance or phosphorylation status of storage proteins during seed germination have been monitored using 2-DE-based reference maps [29,128,131,132]. Ghelis et al. [128] reported that the status of Tyr phosphorylation for several cruciferin precursors and cruciferin subunits in Arabidopsis seeds was modulated in response to ABA using 2-DE-based maps and the identification of phosphorylated Tyr residues by means of anti-phosphotyrosine antibodies in western blots. It was found that cruciferins treated with ABA exhibited higher phosphorylation levels than control seeds. In rice, Han et al. [131] detected that the highest level of phosphorylation of cupins coincided with the late stage of germination and protein degradation by means of 2-DE combined with Pro-Q DPS staining and MALDI-TOF/TOF MS. Using DIGE-based maps, Dong et al. [132] detected an increased abundance of phosphorylated wheat globulin 3 at 12 h after imbibition. In common beans, the analysis of targeted 2-DE-based phaseolin profiles coupled to protein dephosphorylation with HF-P revealed changes in the phosphorylation status during dry-to-germinating seed transition [29]. Changes in the phosphorylation status unexplained by parallel variations in the amount of protein are suggestive of their functional role [96]. Importantly, highly phosphorylated phaseolin isoforms were preferentially degraded in germinating seeds. These results support the conclusion that phosphorylation-dependent degradation plays a significant role in the mobilization of phaseolin. It has been suggested that phosphorylation can cause conformational changes in the protein and promote its mobilization during germination [127]. Overall, the molecular pathways, phosphorylation sites and specific kinases/phosphatases governing variations in phosphorylation status are totally unknown.

Protein glycosylation is involved in the modulation of relevant biological processes such as protein folding, protein stability, protein-protein interactions and interaction with membrane components [147–149]. Asparagine (N)-linked glycosylation is the major co- and post-translational modification of proteins in plants [150]. The application of a great diversity of molecular techniques permitted the identification of glycosylated isoforms in many types of storage proteins and species: globulins (7S- and 11S-globulins and convicilin) in adzuki bean (*Vigna angularis* L.), blue lupins (*Lupinus angustifolius* L.), cocoa beans (*Theobroma cacao* L.), common beans, hazelnuts (*Corylus avellane* L.), lentils, Lotus (*Lotus japonicus* L.), mung beans (*Vigna radiata* L.), peas (*Pisum sativum* L.), peanuts (*Arachis hipogea* L.), soybeans and white lupin (*Lupinus albus* L.) [42,120,121,151–179]; prolamins (γ3-hordein) in barley [180]; albumins (3S albumins) in Inca peanuts (*Plukenetia volubilis* L.) [181]; glutelins in rice [135,182]; VSPs (patatin) in potatoes [30,43,183–186]; and lectins (monocot mannose-binding lectin, phytohemagglutinin) in air potatoes (*Dioscorea bulbifera* L.), common beans and lotus [121,187–193].

Identification and profiling studies of glycosylated storage proteins using 2-DE-based maps together with various other techniques are listed in Table 3. Most of these studies are addressed to the identification of glycosylated isoforms, the assessment of differential degrees of glycosylation and effects in food allergy. The biological role of glycosylated forms remains largely unknown. Interestingly, Santos et al. [177] reported that the glucoside hydrolase β-*N*-acetylhexosaminidase (β-NAHase) is involved in α-conglutin mobilization in white lupin storage proteins.

Table 2. List of 2-DE-based seed phosphoproteomic studies including storage proteins.

Storage Protein Type	Storage Protein Subtype	Seed Stage	Additional Techniques	Species	References
Globulin	12S cruciferin	Development	Pro-Q DPS LC-MS/MS	Rapeseed (*Brassica napus* L.)	[113]
	12S triticin Globulin 3	Development	Pro-Q DPS MALDI-TOF MALDI-TOF/TOF	Wheat (*Triticum aestivum* L.)	[67]
	12S cruciferin	Dormancy	1-DE, Pro-Q DPS immunoblotting LC-MS/MS	*Arabidopsis thaliana* L.	[126,127]
	7S phaseolin	Dormancy/Germination	Pro-Q DPS, HF-P MALDI-TOF MALDI-TOF/TOF	Common bean (*Phaseolus vulgaris* L.)	[29]
	12S cruciferin	Germination	Western blotting MALDI-TOF MALDI-TOF/TOF	*Arabidopsis thaliana* L.	[128]
	Cupin	Germination	Pro-Q DPS MALDI-TOF/TOF	Rice (*Oryza sativa* L.)	[131]
	Globulin 3	Germination	Pro-Q DPS LC-MS/MS	Wheat (*Triticum aestivum* L.)	[132]
Albumin	2S napin	Dormancy	1-DE, Pro-Q DPS immunoblotting LC-MS/MS	*Arabidopsis thaliana* L.	[126,127]
Glutelin	N/A	Development	Pro-Q DPS LC-MS/MS	Rice (*Oryza sativa* L.)	[135]
Vegetative	Patatin	Dormancy	Pro-Q DPS, HF-P MALDI-TOF MALDI-TOF/TOF	Potato (*Solanum tuberosum* L.)	[30]

N/A, not available.

Table 3. List of 2-DE-based seed glycoproteomic studies including storage proteins.

Storage Protein Type	Storage Protein Subtype	Additional Techniques	Species	References
Globulin	7S vicilin	1-DE, Glycoprotein staining	Cocoa bean (*Theobroma cacao* L.)	[179]
	7S phaseolin	1-DE, Fluorography Radioactive labelling of sugars, Concanavalin A binding Immunoaffinity chromatography N-deglycosylation	Common bean (*Phaseolus vulgaris* L.)	[42,161,162]
	7S convicilin	N-deglycosylation	Lotus (*Lotus japonicus* L.)	[121]
Glutelin	N/A	Glycoprotein staining, LC-MS/MS	Rice (*Oryza sativa* L.)	[135]
Lectin	N/A	N-deglycosylation	Lotus (*Lotus japonicus* L.)	[121]
Vegetative	Patatin	N-deglycosylation MALDI-TOF, MALDI-TOF/TOF	Potato (*Solanum tuberosum* L.)	[30,43,184]

N/A, not available.

5. Application Areas in Seed Breeding

5.1. Seed Quality

Seed protein quality is an essential trait in seed breeding programs. The nutritional quality of proteins is largely dependent on their essential amino acid (EAA) composition, total protein content and digestibility. Seed proteins are often deficient in specific EAA such as lysine, tryptophan, threonine and methionine. For example, high relative concentrations of lysine can be found in potato tuber but it is a nutritionally limiting EAA in most cereals [194]; whereas soybeans and common beans are deficient in methionine [194,195]. Storage proteins are abundant and determine to a great extent seed protein quality. For example, the relative abundance of prolamins in cereals has a key influence for protein quality because of their deficiences in EAA [73]. In particular, zein is a prolamin that accounts for between 50 and 70% of the total seed protein of maize and is mainly deficient in the content of lysine and tryptophan followed by methionine [61,196,197]. The particular mix of abundant storage proteins can also determine the final quality of seed proteins. For example, glycinin (11S legumin type) and conglycinin (7S vicilin type) are the two major soybean storage proteins, but glycinin harbors three to four times more sulfur-containing amino acids than conglycinin [198].

2-DE-based maps are a very effective tool for screening and selecting varieties containing specific protein storage isoforms linked to high protein quality in plant breeding. This proteomic approach has been addressed in a variety of crops. For instance, wild rice species are a valuable source of genetic resources for improving the nutritional quality of rice by increasing the glutelin content to the detriment of prolamins [199,200]. The comparison of 2-DE-based maps between wild rice species and rice cultivars revealed new subunits and precursors of glutelin in wild rice species [199]. 2-DE gels also revealed that the content of glutelins in an ancient Chinese wild rice (*Zizania latifolia* (Griseb.) *Turcz.*) was approximately twice as high as that of the Indica rice cultivar [200]. Zarkadas et al. [195] also reported great variability among soybean cultivars for glycinin and β-conglycinin using 2-DE. In common bean, López-Pedrouso et al. [44] reported that pairwise proteomic distances estimated from wild and domesticated accessions of the major Mesoamerican and Andean gene pools assessed by targeted 2-DE of the phaseolin provide valuable information for identifying outlier cultivars with increased content in methionine.

A number of factors modeling the genetic structure of populations can generate and/or maintain genome-wide non-random associations between alleles at different loci (linkage or gametic disequilibrium) such as founder effects, bottlenecks, inbreeding and selection [201]. These factors or combination of factors often operate in plant breeding. Accordingly, storage proteins encoded by multigene families can be used to detect nonrandom associated quantitative trait loci (QTLs) underlying quality traits. In this regard, the nutritional quality of protein and the starch content and average weight of potato tubers were found to be correlated with patatin content [202,203].

Different types of transgenic-based strategies have been addressed at the improvement of seed protein quality from storage proteins. Some strategies rely on the ectopic expression of transgenes coding for high quality proteins that correct seed deficiencies in the amino acid composition of storage proteins. Shekhar et al. [50] introduced the seed albumin gene *AmA1* from *Amaranthus hypochondriacus* into sweet potato (*Ipomoea batatas* L.) by Agrobacterium-mediated transformation to assess the behavior of storage proteins in a non-native system. *AmA1* is rich in all EAA whereas sweet potato proteins are deficient in tryptophan and sulfur-containing amino acids. Comparative proteomics revealed that 2-DE profiles of transgenic tubers exhibited a higher number of protein spots than wild-type tubers. The results suggest that overexpression of *AmA1* in sweet potato tubers seems to have a marked effect on nutrient acquisition, which facilitates an increase in the overall protein and amino acid content. Other alternative transgenics-based approaches are used to overproduce one particular seed protein with higher nutritional quality than the remaining set of storage proteins. For example, the overexpression of glycinin enables an increase in sulfur amino acids in soybean seeds, taking into account that the content of glycinin correlates negatively to the content of β-conglycinin [204].

El-Shemy et al. [198] transformed soybean embryos with a chimeric proglycining gene encoding a methionine-rich glycinin. The comparison of transgenic and untransformed soybean lines by 2-DE revealed an increased accumulation of glycinin in transgenic soybeans.

5.2. Gluten Disorders and Allergies

Gluten proteins and gluten-like proteins are the main factor triggering coeliac disease (CD), non-coeliac gluten sensitivity and gluten allergies in genetically susceptible individuals [51,205,206]. CD is an autoimmune condition caused by human intolerance to wheat gluten and related proteins from rye (secalins, *Secale cereale* L.), barley (hordeins) and oat (avenins, *Avena sativa* L.) that primarily affect the small intestine [72,206]. Gluten is composed of a combination of two toxic prolamines in CD, glutenins and gliadins, but gliadins contain most of the epitopes triggering CD [51,72]. A gluten-free diet is often low in fiber and minerals, high in sucrose and saturated fatty acids, and more expensive [207,208]. A wide-variety of strategies have been applied for the selection and breeding of less toxic varieties. These include obtaining varieties with a lower dose or a different composition of gluten proteins. García-Molina et al. [51] carried out a 2-DE-based proteomic study to evaluate the effects of the strong down-regulation of gliadins on the expression of target and non-target proteins. For this purpose, transgenic wheat lines with downregulation of gliadin expression were obtained by RNA interference (RNAi) technology. As expected, transgenic lines showed a lower abundance of gliadins with respect to control lines. However, the glutelin fraction and other allergen-related wheat proteins increased in low-gliadin lines by a compensation effect. Kawaura et al. [209] obtained aneuploid wheat lines to reduce CD immunotoxicity in breeding programs. An analysis of 2-DE profiles disclosed that α-gliadins containing major CD epitopes were lost in tetrasomic lines. In barley, Tanner et al. [206] obtained an ultra-low gluten variety (hordein content below 5 ppm) by combining three recessive alleles with potential application in the preparation of foods and beverages for CD patients and people who cannot tolerate gluten. Only reduced amounts of the γ-3-hordein protein were observed in the ultra-low gluten variety by 2-DE, in accordance with other protein quantitative determinations. 2-DE also contributed to demonstrating that wheat α-gliadins can be compensated by the addition of avenins to the floor to improve dough quality, taking into account that a minority of CD patients are sensitive to oat avenins [210]. Rizzello et al. [211] showed by in vitro analysis that making bread from flour with an intermediate content of gluten improves its digestibility and nutritional quality without the loss of the chemical, structural and sensory characteristics of traditional breads. 2-DE revealed increased protein degradation in flour with an intermediate content of gluten during fermentation. The authors suggested that this wheat product might be useful to prevent, delay or treat susceptibility to gluten sensitivity, a gluten reaction that does not involve allergic or autoimmune mechanisms.

5.3. Seed Longevity

Dry seed longevity is an essential complex trait for the biodiversity conservation of cultivated plants. Seed longevity and the germination vigor rate slowly decrease during storage ageing, influenced by abiotic and biotic variables, including storage conditions (e.g., temperature and humidity) and genetic factors [212–214]. Compelling evidence indicates that antioxidant systems (antioxidative enzymes and antioxidants) deteriorate during seed ageing leading to the accumulation of reactive oxygen species (ROS) and oxidative damage [49,124,215]. SSPs undergo extensive oxidization (often carbonylation) during long-term seed storage due to their abundance and high affinity to oxidation [49,215–217]. Seed ageing profiling in rice assessed by 2-DE followed by western blotting with antidinitrophenyl hydrazone antibodies revealed that carbonylated SSPs accumulate at the critical node of seed ageing leading to a rapid decline in seed viability [124]. Nguyen et al. [49] proposed that SSPs may be buffers for seed oxidative stress, able to protect relevant proteins for seed germination and seedling development from proteomic profile analysis of Arabidopsis cruciferin mutants based on

2-DE and LC-MS/MS. Dobiesz et al. [214] reported that β- and δ-conglutins may be a useful biomarker of lupin (*Lupinus luteus* L.) seed viability during long-term storage using 2-DE and LC-MS/MS.

5.4. Other Applications

The analysis of storage proteins by 2-DE-based maps has also contributed to the development of other application areas such as antifungal, antibacterial and insect susceptibility [45–47,218], the identification of allergens [46], drought stress [48,219], wheat cultivar identification in blended flour [220] and the large-scale production of therapeutic proteins [221].

6. General Conclusions and Perspectives

This review shows that the use of 2-DE combined with MS is of vital importance not only to advancing the knowledge of the isoforms of storage proteins and their dynamic changes during seed development and germination in a wide diversity of plants, but also in relevant fields closely connected to seed breeding. Therefore, the employment of 2-DE is expected to follow over the next years due to its high efficiency in the characterization of storage proteins across different biological scenarios. Gel-based and shotgun proteomics are alternative strategies for proteome analysis that have advantages and limitations but complement each other. The joint use of gel-based and gel-free methodologies will probably continue to be necessary in follow-up studies to understand the complex biology of storage proteins. Despite significant progress over the last decades, proteomics faces major challenges in the coming years to unravel the complex molecular puzzle of regulatory networks underlying the activities, functions, and interactions of storage proteins over the lifecycle of seeds. In particular, further experiments are clearly needed to assess the exact role of phosphorylated isoforms and specific phosphorylation sites during seed development and germination. This huge task will probably require the integration of multi-omics data with the help of new bioinformatic tools.

Author Contributions: C.Z. conceived, coordinated and supervised the study. All authors (D.M., J.B., M.L.-P., D.F. and C.Z.) wrote, revised and approved the final manuscript.

Funding: This research received no external funding.

Acknowledgments: We thank two anonymous reviewers for their valuable comments and suggestions to improve the quality of the paper.

Conflicts of Interest: The authors declare no conflict of interest.

Abbreviations

1-DE	One-dimensional electrophoresis
2-DE	Two-dimensional electrophoresis
ABA	Abscisic acid
CD	Coeliac disease
DIGE	Difference gel electrophoresis
EAA	Essential amino acid
GA	Gibberellic acid
HF-P	Hydrogen fluoride-pyridine
IMAC	Immobilized metal affinity chromatography
M_r	Relative molecular mass
MS	Mass spectrometry
pI	Isoelectric point
PR	Phosphorylation rate
Pro-Q DPS	Pro-Q Diamond phosphoprotein stain
PTM	Post-translational modification
SSP	Seed storage protein
VSP	Vegetative storage protein

References

1. Shewry, P.R.; Napier, J.A.; Tatham, A.S. Seed storage proteins: Structures and biosynthesis. *Plant Cell* **1995**, *7*, 945–956. [CrossRef] [PubMed]
2. Müntz, K. Deposition of storage proteins. *Plant Mol. Biol.* **1998**, *38*, 77–99. [CrossRef] [PubMed]
3. Shewry, P.R.; Halford, N.G. Cereal seed storage proteins: Structures, properties and role in grain utilization. *J. Exp. Bot.* **2002**, *53*, 947–958. [CrossRef] [PubMed]
4. Gallardo, K.; Firnhaber, C.; Zuber, H.; Héricher, D.; Belghazi, M.; Henry, C.; Küster, H.; Thompson, R. A combined proteome and transcriptome analysis of developing *Medicago truncatula* seeds evidence for metabolic specialization of maternal and filial tissues. *Mol. Cell. Proteom.* **2007**, *6*, 2165–2179. [CrossRef] [PubMed]
5. Tan-Wilson, A.L.; Wilson, K.A. Mobilization of seed protein reserves. *Physiol. Plant.* **2012**, *145*, 140–153. [CrossRef] [PubMed]
6. van Vliet, S.; Burd, N.; van Loon, L. The skeletal muscle anabolic response to plant-versus animal-based protein consumption. *J. Nutr.* **2015**, *145*, 1981–1991. [CrossRef] [PubMed]
7. Pasiakos, S.; Agarwal, S.; Lieberman, H.; Fulgoni, V. Sources and amounts of animal, dairy, and plant protein intake of US adults in 2007–2010. *Nutrients* **2015**, *7*, 7058–7069. [CrossRef] [PubMed]
8. FAOSTAT. Statistics Division of the FAO. Available online: http://www.fao.org/faostat/en/ (accessed on 22 June 2018).
9. World Bank List of Economies. Available online: http://www.worldbank.org// (accessed on 22 June 2018).
10. Aguirrezábal, L.; Martre, P.; Pereyra-Irujo, G.; Echarte, M.M.; Izquierdo, N. Improving grain quality: Ecophysiological and modeling tools to develop management and breeding strategies. In *Crop Physiology, Applications for Genetic Improvement and Agronomy*, 2nd ed.; Sadras, V., Calderini, D., Eds.; Academic Press: London, UK, 2015; pp. 423–465. ISBN 9780124171046.
11. Racusen, D. Lipid acyl hydrolase of patatin. *Can. J. Bot.* **1984**, *62*, 1640–1644. [CrossRef]
12. Liu, Y.W.; Han, C.H.; Lee, M.H.; Hsu, F.L.; Hou, W.C. Patatin, the tuber storage protein of potato (*Solanum tuberosum* L.), exhibits antioxidant activity in vitro. *J. Agric. Food Chem.* **2003**, *51*, 4389–4393. [CrossRef] [PubMed]
13. de Souza Cândido, E.; Pinto, M.F.; Pelegrini, P.B.; Lima, T.B.; Silva, O.N.; Pogue, R.; Grossi-de-Sá, M.F.; Franco, O.L. Plant storage proteins with antimicrobial activity: Novel insights into plant defense mechanisms. *FASEB J.* **2011**, *25*, 3290–3305. [CrossRef] [PubMed]
14. Joshi, J.; Panduranga, S.; Diapari, M.; Marsolais, F. Comparison of Gene Families: Seed Storage and Other Seed Proteins. In *The Common Bean Genome*; de la Vega, M.P., Santalla, M., Marsolais, F., Eds.; Springer: Cham, Switzerland, 2017; pp. 201–219. ISBN 978-3-319-63524-8.
15. Girke, T.; Todd, J.; Ruuska, S.; White, J.; Benning, C.; Ohlrogge, J. Microarray analysis of developing *Arabidopsis* seeds. *Plant Physiol.* **2000**, *12*, 1570–1581. [CrossRef]
16. Tzafrir, I.; Dickerman, A.; Brazhnik, O.; Nguyen, Q.; McElver, J.; Frye, C.; Patton, D.; Meinke, D. The *Arabidopsis* seedgenes project. *Nucleic Acids Res.* **2003**, *31*, 90–93. [CrossRef] [PubMed]
17. McElver, J.; Tzafrir, I.; Aux, G.; Rogers, R.; Ashby, C.; Smith, K.; Thomas, C.; Schetter, A.; Zhou, Q.; Cushman, M.A.; et al. Insertional mutagenesis of genes required for seed development in *Arabidopsis thaliana*. *Genetics* **2001**, *159*, 1751–1763. [PubMed]
18. Meinke, D.; Muralla, R.; Sweeney, C.; Dickerman, A. Identifying essential genes in *Arabidopsis thaliana*. *Trends Plant Sci.* **2008**, *13*, 483–491. [CrossRef] [PubMed]
19. Le, B.H.; Cheng, C.; Bui, A.Q.; Wagmaister, J.A.; Henry, K.F.; Pelletier, J.; Kwong, L.; Belmonte, M.; Kirkbride, R.; Horvath, S.; et al. Global analysis of gene activity during *Arabidopsis* seed development and identification of seed-specific transcription factors. *Proc. Natl. Acad. Sci. USA* **2010**, *107*, 8063–8070. [CrossRef] [PubMed]
20. Gacek, K.; Bartkowiak-Broda, I.; Batley, J. Genetic and molecular regulation of Seed Storage Proteins (SSPs) to improve protein nutritional value of oilseed rape (*Brassica napus* L.) seeds. *Front. Plant Sci.* **2018**, *9*, 890. [CrossRef] [PubMed]
21. Rasheed, A.; Xia, X.; Yan, Y.; Appels, R.; Mahmood, T.; He, Z. Wheat seed storage proteins: Advances in molecular genetics, diversity and breeding applications. *J. Cereal Sci.* **2014**, *60*, 11–24. [CrossRef]

22. Finch-Savage, W.; Leubner-Metzger, G. Seed dormancy and the control of germination. *New Phytol.* **2006**, *171*, 501–523. [CrossRef] [PubMed]

23. Hirayama, T.; Shinozaki, K. Perception and transduction of abscisic acid signals: Keys to the function of the versatile plant hormone ABA. *Trends Plant Sci.* **2007**, *12*, 343–351. [CrossRef] [PubMed]

24. Gutierrez, L.; Van Wuytswinkel, O.; Castelain, M.; Bellini, C. Combined networks regulating seed maturation. *Trends Plant Sci.* **2007**, *12*, 294–300. [CrossRef] [PubMed]

25. Han, C.; Yang, P. Studies on the molecular mechanisms of seed germination. *Proteomics* **2015**, *15*, 1671–1679. [CrossRef] [PubMed]

26. Née, G.; Kramer, K.; Nakabayashi, K.; Yuan, B.; Xiang, Y.; Miatton, E.; Finkemeier, I.; Soppe, W.J.J. Delay of germination1 requires PP2C phosphatases of the ABA signalling pathway to control seed dormancy. *Nat. Commun.* **2017**, *8*, 72. [CrossRef] [PubMed]

27. Kim, H.T.; Choi, U.K.; Ryu, H.S.; Lee, S.J.; Kwon, O.S. Mobilization of storage proteins in soybean seed (*Glycine max* L.) during germination and seedling growth. *Biochim. Biophys. Acta* **2011**, *1814*, 1178–1187. [CrossRef] [PubMed]

28. The Potato Genome Sequencing Consortium. Genome sequence and analysis of the tuber crop potato. *Nature* **2011**, *475*, 189–195. [CrossRef] [PubMed]

29. López-Pedrouso, M.; Alonso, J.; Zapata, C. Evidence for phosphorylation of the major seed storage protein of the common bean and its phosphorylation-dependent degradation during germination. *Plant Mol. Biol.* **2014**, *84*, 415–428. [CrossRef] [PubMed]

30. Bernal, J.; López-Pedrouso, M.; Franco, D.; Bravo, S.; García, L.; Zapata, C. Identification and mapping of phosphorylated isoforms of the major storage protein of potato based on two-dimensional electrophoresis. In *Advances in Seed Biology*; Jimenez-Lopez, J.C., Ed.; InTech: Rijeka, Croatia, 2017; pp. 65–82. ISBN 978-953-51-3621-7.

31. Jorrín, J.V.; Maldonado, A.M.; Castillejo, M.A. Plant proteome analysis: A 2006 update. *Proteomics* **2007**, *7*, 2947–2962. [CrossRef] [PubMed]

32. Kersten, B.; Agrawal, G.K.; Durek, P.; Neigenfind, J.; Schulze, W.; Walther, D.; Rakwal, R. Plant phosphoproteomics: An update. *Proteomics* **2009**, *9*, 964–988. [CrossRef] [PubMed]

33. Silva-Sanchez, C.; Li, H.; Chen, S. Recent advances and challenges in plant phosphoproteomics. *Proteomics* **2015**, *15*, 1127–1141. [CrossRef] [PubMed]

34. Miernyk, J.A.; Hajduch, M. Seed proteomics. *J. Proteom.* **2011**, *74*, 389–400. [CrossRef] [PubMed]

35. Miernyk, J.A. Seed Proteomics. In *Plant Proteomics*; Jorrin-Novo, J.J., Komatsu, S., Weckwerth, W., Wienkoop, S., Eds.; Humana Press: New York, NY, USA, 2014; pp. 361–379. ISBN 978-1-62703-630-6.

36. Narula, K.; Sinha, A.; Haider, T.; Chakraborty, N.; Chakraborty, S. Seed Proteomics: An Overview. In *Agricultural Porteomics*; Salekdeh, G.H., Ed.; Springer: Cham, Switzerland, 2016; Volume 1, pp. 31–53. ISBN 978-3-319-43273-1.

37. Zargar, S.M.; Mahajan, R.; Nazir, M.; Nagar, P.; Kim, S.T.; Rai, V.; Masi, A.; Ahmad, S.M.; Shah, R.A.; Ganai, N.A.; et al. Common bean proteomics: Present status and future strategies. *J. Proteom.* **2017**, *169*, 239–248. [CrossRef] [PubMed]

38. O'Farrel, P.H. High resolution two-dimensional electrophoresis of proteins. *J. Biol. Chem.* **1975**, *250*, 4007–4021.

39. Barbier-Brygoo, H.; Joyard, J. Focus on plant proteomics. *Plant Physiol. Biochem.* **2004**, *42*, 913–917. [CrossRef] [PubMed]

40. Chen, S.; Harmon, A.C. Advances in plant proteomics. *Proteomics* **2006**, *6*, 5504–5516. [CrossRef] [PubMed]

41. Ghatak, A.; Chaturvedi, P.; Weckwerth, W. Cereal crop proteomics: Systemic analysis of crop drought stress responses towards marker-assisted selection breeding. *Front. Plant Sci.* **2017**, *8*, 757. [CrossRef] [PubMed]

42. de la Fuente, M.; López-Pedrouso, M.; Alonso, J.; Santalla, M.; de Ron, A.M.; Alvarez, G.; Zapata, C. In-depth characterization of the phaseolin protein diversity of common bean (*Phaseolus vulgaris* L.) based on two-dimensional electrophoresis and mass spectrometry. *Food Technol. Biotechnol.* **2012**, *50*, 315–325.

43. Bárta, J.; Bártová, V.; Zdráhal, Z.; Šedo, O. Cultivar variability of patatin biochemical characteristics: Table versus processing potatoes (*Solanum tuberosum* L). *J. Agric. Food Chem.* **2012**, *60*, 4369–4378. [CrossRef] [PubMed]

44. López-Pedrouso, M.; Bernal, J.; Franco, D.; Zapata, C. Evaluating two-dimensional electrophoresis profiles of the protein phaseolin as markers of genetic differentiation and seed protein quality in common bean (*Phaseolus vulgaris* L.). *J. Agric. Food Chem.* **2014**, *62*, 7200–7208. [CrossRef] [PubMed]

45. Flores, T.; Alape-Girón, A.; Flores-Díaz, M.; Flores, H.E. Ocatin. A novel tuber storage protein from the andean tuber crop oca with antibacterial and antifungal activities. *Plant Physiol.* **2002**, *128*, 1291–1302. [CrossRef] [PubMed]

46. Palomares, O.; Cuesta-Herranz, J.; Vereda, A.; Sirvent, S.; Villalba, M.; Rodríguez, R. Isolation and identification of an 11S globulin as a new major allergen in mustard seeds. *Ann. Allergy Asthma Immunol.* **2005**, *94*, 586–592. [CrossRef]

47. Collins, R.M.; Afzal, M.; Ward, D.A.; Prescott, M.C.; Sait, S.M.; Rees, H.H.; Tomsett, A.B. Differential proteomic analysis of *Arabidopsis thaliana* genotypes exhibiting resistance or susceptibility to the insect herbivore, *Plutella xylostella*. *PLoS ONE* **2010**, *5*, e10103. [CrossRef] [PubMed]

48. Zhang, Y.F.; Huang, X.W.; Wang, L.L.; Wei, L.; Wu, Z.H.; You, M.S.; Li, B.Y. Proteomic analysis of wheat seed in response to drought stress. *J. Integr. Agric.* **2014**, *13*, 919–925. [CrossRef]

49. Nguyen, T.P.; Cueff, G.; Hegedus, D.D.; Rajjou, L.; Bentsink, L. A role for seed storage proteins in *Arabidopsis* seed longevity. *J. Exp. Bot.* **2015**, *66*, 6399–6413. [CrossRef] [PubMed]

50. Shekhar, S.; Agrawal, L.; Mishra, D.; Buragohain, A.K.; Unnikrishnan, M.; Mohan, C.; Chakraborty, S.; Chakraborty, N. Ectopic expression of amaranth seed storage albumin modulates photoassimilate transport and nutrient acquisition in sweetpotato. *Sci. Rep.* **2016**, *6*, 25384. [CrossRef] [PubMed]

51. García-Molina, M.D.; Muccilli, V.; Saletti, R.; Foti, S.; Masci, S.; Barro, F. Comparative proteomic analysis of two transgenic low-gliadin wheat lines and non-transgenic wheat control. *J. Proteom.* **2017**, *165*, 102–112. [CrossRef] [PubMed]

52. Osborne, T.B. *The Vegetable Proteins*, 2nd ed.; Longmans, Green and Co.: London, UK, 1924; pp. 1–154.

53. Marla, S.; Bharatiya, D.; Bala, M.; Singh, V.; Kumar, A. Classification of rice seed storage proteins using neural networks. *J. Plant Biochem. Biotechnol.* **2010**, *19*, 123–126. [CrossRef]

54. Radhika, V.; Rao, V.S. Computational approaches for the classification of seed storage proteins. *J. Food Sci. Technol.* **2015**, *52*, 4246–4255. [CrossRef] [PubMed]

55. Beardmore, T.; Wetzel, S.; Burgess, D.; Charest, P.J. Characterization of seed storage proteins in *Populus* and their homology with *Populus* vegetative storage proteins. *Tree Physiol.* **1996**, *16*, 833–840. [CrossRef] [PubMed]

56. Fujiwara, T.; Nambara, E.; Yamagishi, K.; Goto, D.B.; Naito, S. Storage proteins. *Arabidopsis Book* **2002**, *1*, e0020. [CrossRef] [PubMed]

57. Pikaard, C.S.; Brusca, J.S.; Hannapel, D.J.; Park, W.D. The two classes of genes for the major potato tuber protein, patatin, are differentially expressed in tubers and roots. *Nucleic Acids Res.* **1979**. [CrossRef]

58. Mignery, G.A.; Pikaard, C.; Park, W. Molecular characterization of the patatin multigene family of potato. *Gene* **1988**, *62*, 27–44. [CrossRef]

59. Staswick, P.E. Novel regulation of vegetative storage protein genes. *Plant Cell* **1990**, *2*, 1–6. [CrossRef] [PubMed]

60. Consoli, L.; Damerval, C. Quantification of individual zein isoforms resolved by two-dimensional electrophoresis: Genetic variability in 45 maize inbred lines. *Electrophoresis* **2001**, *22*, 2983–2989. [CrossRef]

61. Lund, G.; Ciceri, P.; Viotti, A. Maternal-specific demethylation and expression of specific alleles of zein genes in the endosperm of *Zea mays* L. *Plant J.* **1995**, *8*, 571–581. [CrossRef] [PubMed]

62. Pinheiro, C.; Sergeant, K.; Machado, C.M.; Renaut, J.; Ricardo, C.P. Two traditional maize inbred lines of contrasting technological abilities are discriminated by the seed flour proteome. *J. Proteome Res.* **2013**, *12*, 3152–3165. [CrossRef] [PubMed]

63. Xu, J.H.; Messing, J. Organization of the prolamin gene family provides insight into the evolution of the maize genome and gene duplications in grass species. *Proc. Natl. Acad. Sci. USA* **2008**, *105*, 14330–14335. [CrossRef] [PubMed]

64. Ning, F.; Niu, L.; Yang, H.; Wu, X.; Wang, W. Accumulation profiles of embryonic salt-soluble proteins in maize hybrids and parental lines indicate matroclinous inheritance: A proteomic analysis. *Front. Plant Sci.* **2017**, *8*, 1824. [CrossRef] [PubMed]

65. Payne, P.I. Genetics of wheat storage proteins and the effect of allelic variation on bread-making quality. *Ann. Rev. Plant Physiol.* **1987**, *38*, 141–153. [CrossRef]

66. Shewry, P.R.; Tatham, A.S. The prolamin storage proteins of cereal seeds: Structure and evolution. *Biochem. J.* **1990**, *267*, 1–12. [CrossRef] [PubMed]

67. Guo, G.; Lv, D.; Yan, X.; Subburaj, S.; Ge, P.; Li, X.; Hu, Y.; Yan, Y. Proteome characterization of developing grains in bread wheat cultivars (*Triticum aestivum* L.). *BMC Plant Biol.* **2012**, *12*, 147. [CrossRef] [PubMed]

68. Malik, A.H. Nutrient uptake, transport and translocation in cereals: Influences of environmental and farming conditions. *Swed. Univ. Agric. Sci.* **2009**, *1*, 1–46.

69. Zhou, J.; Liu, D.; Deng, X.; Zhen, S.; Wang, Z.; Yan, Y. Effects of water deficit on breadmaking quality and storage protein compositions in bread wheat (*Triticum aestivum* L.). *J. Sci. Food Agric.* **2018**, *98*, 4357–4368. [CrossRef] [PubMed]

70. Xie, Z.; Wang, C.; Wang, K.; Wang, S.; Li, X.; Zhang, Z.; Ma, W.; Yan, Y. Molecular characterization of the celiac disease epitope domains in α-gliadin genes in *Aegilops tauschii* and hexaploid wheats (*Triticum aestivum* L.). *Theor. Appl. Genet.* **2010**, *121*, 1239–1251. [CrossRef] [PubMed]

71. Cavazos, A.; Gonzalez de Mejia, E. Identification of bioactive peptides from cereal storage proteins and their potential role in prevention of chronic diseases. *Compr. Rev. Food Sci. Food Saf.* **2013**, *12*, 364–380. [CrossRef]

72. Ferranti, P.; Mamone, G.; Picariello, G.; Addeo, F. Mass spectrometry analysis of gliadins in celiac disease. *J. Mass Spectrom.* **2007**, *42*, 1531–1548. [CrossRef] [PubMed]

73. Yadav, D.; Singh, N. Wheat triticin: A potential target for nutritional quality improvement. *Asian J. Biotechnol.* **2011**, *3*, 1–21. [CrossRef]

74. Zhang, W.; Sun, J.; Zhao, G.; Wang, J.; Liu, H.; Zheng, H.; Zhao, H.; Zou, D. Association analysis of the glutelin synthesis genes *GluA* and *GluB1* in a *Japonica* rice collection. *Mol. Breed.* **2017**, *37*, 129. [CrossRef]

75. Kim, H.J.; Lee, J.Y.; Yoon, U.H.; Lim, S.H.; Kim, Y.M. Effects of reduced prolamin on seed storage protein composition and the nutritional quality of rice. *Int. J. Mol. Sci.* **2013**, *14*, 17073–17084. [CrossRef] [PubMed]

76. He, Y.; Wang, S.; Ding, Y. Identification of novel glutelin subunits and a comparison of glutelin composition between *japonica* and *indica* rice (*Oryza sativa* L.). *J. Cereal Sci.* **2013**, *57*, 362–371. [CrossRef]

77. Bártová, V.; Bárta, J. Chemical composition and nutritional value of protein concentrates isolated from potato (*Solanum tuberosum* L.) fruit juice by precipitation with ethanol or ferric chloride. *J. Agric. Food Chem.* **2009**, *57*, 9028–9034. [CrossRef] [PubMed]

78. Jørgensen, M.; Stensballe, A.; Welinder, K.G. Extensive post-translational processing of potato tuber storage proteins and vacuolar targeting. *FEBS J.* **2011**, *278*, 4070–4087. [CrossRef] [PubMed]

79. Boehm, J.D.; Nguyen, V.; Tashiro, R.M.; Anderson, D.; Shi, C.; Wu, X.; Woodrow, L.; Yu, K.; Cui, Y.; Li, Z. Genetic mapping and validation of the loci controlling 7S α′ and 11S A-type storage protein subunits in soybean [*Glycine max* (L.) Merr.]. *Theor. Appl. Genet.* **2018**, *131*, 659–671. [CrossRef] [PubMed]

80. Goyal, R.; Sharma, S. Genotypic variability in seed storage protein quality and fatty acid Composition of soybean [*Glycine max* (L.) Merrill]. *Legum. Res.* **2015**, *38*, 297–302. [CrossRef]

81. Friedman, M.; Brandon, D.L. Nutritional and health benefits of soy proteins. *J. Agric. Food Chem.* **2001**, *49*, 1069–1086. [CrossRef] [PubMed]

82. Silva, F.; Nogueira, L.C.; Gonçalves, C.; Ferreira, A.A.; Ferreira, I.M.P.L.V.O.; Teixeira, N. Electrophoretic and HPLC methods for comparative study of the protein fractions of malts, worts and beers produced from Scarlett and Prestige barley (*Hordeum vulgare* L.) varieties. *Food Chem.* **2008**, *106*, 820–829. [CrossRef]

83. Quiroga, I.; Regente, M.; Pagnussat, L.; Maldonado, A.; Jorrín, J.; de la Canal, L. Phosphorylated 11S globulins in sunflower seeds. *Seed Sci. Res.* **2013**, *23*, 199–204. [CrossRef]

84. Youle, R.J.; Huang, A.H.C. Occurrence of low molecular weight and high cysteine containing albumin storage protein in oil-seeds of diverse species. *Am. J. Bot.* **1981**, *68*, 44–48. [CrossRef]

85. Žilić, S.; Barać, M.; Pešić, M.; Crevar, M.; Stanojević, S.; Nišavić, A.; Saratlić, G.; Tolimir, M. Characterization of sunflower seed and kernel proteins. *Helia* **2010**, *33*, 103–113. [CrossRef]

86. Montoya, C.A.; Leterme, P.; Victoria, N.F.; Toro, O.; Souffrant, W.B.; Beebe, S.; Lallès, J.P. Susceptibility of phaseolin to in vitro proteolysis is highly variable across common bean varieties (*Phaseolus vulgaris*). *J. Agric. Food Chem.* **2008**, *56*, 2183–2191. [CrossRef] [PubMed]

87. D'Amico, L.; Valsasina, B.; Daminati, M.G.; Fabbrini, M.S.; Nitti, G.; Bollini, R.; Ceriotti, A.; Vitale, A. Bean homologs of the mammalian glucose regulated proteins: Induction by tunicamycin and interaction with newly synthesized storage proteins in the endoplasmic reticulum. *Plant J.* **1992**, *2*, 443–455. [CrossRef] [PubMed]

88. Mäkienen, O.E.; Sozer, N.; Ercili-Cura, D.; Poutanen, K. Protein form oat: Structure, processes, functionality, and nutrition. In *Sustainable Protein Sources*; Nadathur, S.R., Wanasundara, J.P.D., Scanlin, L., Eds.; Academic Press: London, UK, 2017; pp. 105–119. ISBN 978-0-12-802779-3.

89. Chang, Y.W.; Alli, I.; Konishi, Y.; Ziomek, E. Characterization of protein fractions from chickpea (*Cicer arietinum* L.) and oat (*Avena sativa* L.) seeds using proteomic techniques. *Food Res. Int.* **2011**, *9*, 3094–3104. [CrossRef]

90. Tulbek, M.C.; Lam, R.S.H.; Wang, Y.; Asavajaru, P.; Lam, A. Pea: A sustainbable vegetable protein crop. In *Sustainable Protein Sources*; Nadathur, S.R., Wanasundara, J.P.D., Scanlin, L., Eds.; Academic Press: London, UK, 2017; pp. 145–164. ISBN 978-0-12-802779-3.

91. Barac, M.; Cabrilo, S.; Pesic, M.; Stanojevic, S.; Zilic, S.; Macej, O.; Ristic, N. Profile and Functional Properties of Seed Proteins from Six Pea (*Pisum sativum*) Genotypes. *Int. J. Mol. Sci.* **2010**, *11*, 4973–4990. [CrossRef] [PubMed]

92. Singh, P.K.; Shrivastava, N.; Chaturvedi, K.; Sharma, B.; Bhagyawant, S.S. Characterization of Seed Storage Proteins from Chickpea Using 2D Electrophoresis Coupled with Mass Spectrometry. *Biochem. Res. Int.* **2016**, *12*, 1049462. [CrossRef] [PubMed]

93. Elfalleh, W.; Nasri, N.; Sarraï, N.; Guasmi, F.; Triki, T.; Marzougui, N.; Ferchichi, A. Storage protein contents and morphological characters of some Tunisian pomegranate (*Punica granatum* L.) cultivars. *Acta Bot. Gallica* **2010**, *157*, 401–409. [CrossRef]

94. Scippa, G.S.; Rocco, M.; Ialicicco, M.; Trupiano, D.; Viscosi, V.; Di Michele, M.; Arena, S.; Chiatante, D.; Scaloni, A. The proteome of lentil (*Lens culinaris* Medik.) seeds: Discriminating between landraces. *Electrophoresis* **2010**, *31*, 497–506. [CrossRef] [PubMed]

95. Schatzki, J.; Ecke, W.; Becker, H.C.; Möllers, C. Mapping of QTL for the seed storage proteins cruciferin and napin in a winter oilseed rape doubled haploid population and their inheritance in relation to other seed traits. *Theor. Appl. Genet.* **2014**, *127*, 1213–1222. [CrossRef] [PubMed]

96. Kim, S.G.; Lee, J.S.; Shin, S.H.; Koo, S.C.; Kim, J.T.; Bae, H.H.; Son, B.Y.; Kim, Y.H.; Kim, S.L.; Baek, S.B.; et al. Profiling of differentially expressed proteins in mature kernels of Korean waxy corn cultivars using proteomic analysis. *J. Korean Soc. Appl. Biol. Chem.* **2015**, *58*, 293–303. [CrossRef]

97. Görg, A.; Postel, W.; Günther, S. The current state of two-dimensional electrophoresis with immobilized pH gradients. *Electrophoresis* **1988**, *9*, 531–546. [CrossRef] [PubMed]

98. Görg, A.; Drews, O.; Lück, C.; Weiland, F.; Weiss, W. 2-DE with IPGs. *Electrophoresis* **2009**, *30* (Suppl. 1), 1221–1232. [CrossRef]

99. Weiss, W.; Görg, A. Two-dimensional electrophoresis for plant proteomics. *Methods Mol. Biol.* **2007**, *355*, 121–143. [PubMed]

100. Wheelock, A.M.; Wheelock, C.E. Bioinformatics in gel-based proteomics. In *Plant Proteomics: Technologies, Strategies and Applications*; Agrawal, G.K., Rakwal, R., Eds.; John Wiley & Sons, Inc.: Hoboken, NJ, USA, 2008; pp. 1–18. ISBN 978-0-470-06976-9.

101. Chevalier, F. Highlights on the capacities of "Gel-based" proteomics. *Proteome Sci.* **2010**, *8*, 23. [CrossRef] [PubMed]

102. Rabilloud, T.; Lelong, C. Two-dimensional gel electrophoresis in proteomics: A tutorial. *J. Proteom.* **2011**, *74*, 1829–1841. [CrossRef] [PubMed]

103. Dowsey, A.W.; Morris, J.S.; Gutstein, H.G.; Yang, G.Z. Informatics and statistics for analyzing 2-D gel electrophoresis images. *Methods Mol. Biol.* **2010**, *604*, 239–255. [CrossRef] [PubMed]

104. Görg, A.; Weiss, W.; Dunn, M.J. Current two-dimensional electrophoresis technology for proteomics. *Proteomics* **2004**, *4*, 3665–3685. [CrossRef] [PubMed]

105. Gupta, R.; Min, C.W.; Wang, Y.; Kim, Y.C.; Agrawal, G.K.; Rakwal, R.; Kim, S.T. Expect the unexpected enrichment of "hidden proteome" of seeds and tubers by depletion of storage proteins. *Front. Plant Sci.* **2016**, *7*, 761. [CrossRef] [PubMed]

106. Saravanan, R.S.; Rose, J.K.C. A critical evaluation of sample extraction techniques for enhanced proteomic analysis of recalcitrant plant tissues. *Proteomics* **2004**, *4*, 2522–2532. [CrossRef] [PubMed]

107. Carpentier, S.C.; Witters, E.; Laukens, K.; Deckers, P.; Swennen, R.; Panis, B. Preparation of protein extracts from recalcitrant plant tissues: An evaluation of different methods for two-dimensional gel electrophoresis analysis. *Proteomics* **2006**, *5*, 2497–2507. [CrossRef] [PubMed]

108. Faurobert, M.; Pelpoir, E.; Chaïb, J. Phenol extraction of proteins for proteomic studies of recalcitrant plan tissues. *Methods Mol. Biol.* **2007**, *355*, 9–14. [CrossRef] [PubMed]

109. de la Fuente, M.; Borrajo, A.; Bermúdez, J.; Lores, M.; Alonso, J.; López, M.; Santalla, M.; de Ron, A.M.; Zapata, C.; Alvarez, G. 2-DE-based proteomic analysis of common bean (*Phaseolus vulgaris* L.) seeds. *J. Proteom.* **2011**, *74*, 262–267. [CrossRef] [PubMed]

110. Rabilloud, T.; Chevallet, M.; Luche, S.; Lelong, C. Two-dimensional gel electrophoresis in proteomics: Past, present and future. *J. Proteom.* **2010**, *73*, 2064–2077. [CrossRef] [PubMed]

111. López-Pedrouso, M.; Pérez-Santaescolástica, C.; Franco, D.; Fulladosa, E.; Carballo, J.; Zapata, C.; Lorenzo, J.M. Comparative proteomic profiling of myofibrillar proteins in dry-cured ham with different proteolysis indices and adhesiveness. *Food Chem.* **2018**, *244*, 238–245. [CrossRef] [PubMed]

112. Rabilloud, T. How to use 2D gel electrophoresis in plant proteomics. In *Plant Proteomics: Methods and Protocols*; Jorrin-Novo, J.V., Komatsu, S., Weckwerth, W., Wienkoop, S., Eds.; Humana Press: New York, NY, USA, 2014; pp. 43–50. ISBN 978-1-4939-6029-3.

113. Agrawal, G.K.; Thelen, J.J. Development of a simplified, economical polyacrylamide gel staining protocol for phosphoproteins. *Proteomics* **2005**, *5*, 4684–4688. [CrossRef] [PubMed]

114. Han, C.; Yang, P. Two Dimensional Gel Electrophoresis-Based Plant Phosphoproteomics. *Methods Mol. Biol.* **2016**, *1355*, 213–223. [CrossRef] [PubMed]

115. Kuyama, H.; Toda, C.; Watanabe, M.; Tanaka, K.; Nishimura, O. An efficient chemical method for dephosphorylation of phosphopeptides. *Rapid Commun. Mass Spectrom.* **2003**, *17*, 1493–1496. [CrossRef] [PubMed]

116. Graur, D.; Li, W.H. *Fundamentals of Molecular Evolution. Sinauer Associate*, 2nd ed.; Sinauer Associates: Sunderland, UK, 2000; pp. 304–322. ISBN 9780878932665.

117. Mignery, G.A.; Pikaard, C.S.; Hannapel, D.J.; Park, W.D. Isolation and sequence analysis of cDNAs for the major potato tuber protein, patatin. *Nucleic Acids Res.* **1984**, *12*, 7987–8000. [CrossRef] [PubMed]

118. Gonzalez-Sanchez, M.B.; Lanucara, F.; Helm, M.; Eyers, C.E. Attempting to rewrite history: Challenges with the analysis of histidine-phosphorylated peptides. *Biochem. Soc. Trans.* **2013**, *41*, 1089–1095. [CrossRef] [PubMed]

119. Mehta-D'souza, P. Detection of glycoproteins in polyacrylamide gels using Pro-Q Emerald 300 Dye, a fluorescent periodate schiff-base stain. *Methods Mol. Biol.* **2012**, *869*, 561–566. [CrossRef] [PubMed]

120. Duranti, M.; Scarafoni, A.; Gius, C.; Negri, A.; Faoro, F. Heat-induced synthesis and tunicamycin-sensitive secretion of the putative storage glycoprotein conglutin γ from mature lupin seeds. *Eur. J. Biochem.* **1994**, *222*, 387–393. [CrossRef] [PubMed]

121. Dam, S.; Thaysen-Andersen, M.; Stenkjaer, E.; Lorentzen, A.; Roepstorff, P.; Packer, N.H.; Stougaard, J. Combined N-glycome and N-glycoproteome analysis of the *Lotus japonicus* seed globulin fraction shows conservation of protein structure and glycosylation in legumes. *J. Proteome Res.* **2013**, *12*, 3383–3392. [CrossRef] [PubMed]

122. Borisjuk, L.; Neuberger, T.; Schwender, J.; Heinzel, N.; Sunderhaus, S.; Fuchs, J.; Hay, J.O.; Tschiersch, H.; Braun, H.P.; Denolf, P.; et al. Seed architecture shapes embryo metabolism in oilseed rape. *Plant Cell* **2013**, *25*, 1625–1640. [CrossRef] [PubMed]

123. Friso, G.; van Wijk, K.J. Posttranslational protein modifications in plant metabolism. *Plant Physiol.* **2015**, *169*, 1469–1487. [CrossRef] [PubMed]

124. Yin, X.; Wang, X.; Komatsu, S. Phosphoproteomics: Protein phosphorylation in regulation of seed germination and plant growth. *Curr. Protein Pept. Sci.* **2018**, *19*, 401–412. [CrossRef] [PubMed]

125. Agrawal, G.K.; Thelen, J.J. Large-scale identification and quantitative profiling of phosphoproteins expressed during seed filling in oilseed rape. *Mol. Cell Proteom.* **2006**, *5*, 2044–2059. [CrossRef] [PubMed]

126. Irar, S.; Oliveira, E.; Pagès, M.; Goday, A. Towards the identification of late-embryogenic-abundant phosphoproteome in *Arabidopsis* by 2-DE and MS. *Proteomics* **2006**, *6*, 175–185. [CrossRef] [PubMed]

127. Wan, L.; Ross, A.R.S.; Yang, J.; Hegedus, D.D.; Kermode, A.R. Phosphorylation of the 12 S globulin cruciferin in wild-type and *abi1-1* mutant *Arabidopsis thaliana* (thale cress) seeds. *Biochem. J.* **2007**, *404*, 247–256. [CrossRef] [PubMed]

128. Ghelis, T.; Bolbach, G.; Clodic, G.; Habricot, Y.; Miginiac, E.; Sotta, B.; Jeannette, E. Protein tyrosine kinases and protein tyrosine phosphatases are involved in abscisic acid-dependent processes in *Arabidopsis* seeds and suspension cells. *Plant Physiol.* **2008**, *148*, 1668–1680. [CrossRef] [PubMed]

129. Meyer, L.J.; Gao, J.; Xu, D.; Thelen, J.J. Phosphoproteomic analysis of seed maturation in *Arabidopsis*, rapeseed, and soybean. *Plant Physiol.* **2012**, *159*, 517–528. [CrossRef] [PubMed]

130. Kovaleva, V.; Cramer, R.; Krynytskyy, H.; Gout, I.; Gout, A. Analysis of tyrosine phosphorylation and phosphotyrosine-binding proteins in germinating seeds from Scots pine. *Plant Physiol. Biochem.* **2013**, *67*, 33–40. [CrossRef] [PubMed]

131. Han, C.; Wang, K.; Yang, P. Gel-based comparative phosphoproteomic analysis on rice during germination. *Plant Cell Physiol.* **2014**, *55*, 1376–1394. [CrossRef] [PubMed]

132. Dong, K.; Zhen, S.; Cheng, Z.; Cao, H.; Ge, P.; Yah, Y. Proteomic analysis reveals key proteins and phosphoproteins upon seed germination of wheat (*Triticum aestivum* L.). *Front. Plant Sci.* **2015**, *6*, 1017. [CrossRef] [PubMed]

133. Tilley, K.A.; Schofield, J.D. Detection of phosphotyrosine in the high M_r subunits of wheat glutenin. *J. Cereal Sci.* **1995**, *22*, 17–19. [CrossRef]

134. Facchiano, A.M.; Colonna, G.; Chiosi, E.; Illiano, G.; Spina, A.; Lafiandra, D.; Buonocore, F. In vitro phosphorylation of high molecular weight glutenin subunits from wheat endosperm. *Plant Physiol. Biochem.* **1999**, *37*, 931–938. [CrossRef]

135. Lin, S.K.; Chang, M.C.; Tsai, Y.G.; Lur, H.S. Proteomic analysis of the expression of proteins related to rice quality during caryopsis development and the effect of high temperature on expression. *Proteomics* **2005**, *5*, 2140–2156. [CrossRef] [PubMed]

136. Vilela, B.; Pagès, M.; Riera, M. Emerging roles of protein kinase CK2 in abscisic acid signaling. *Front. Plant Sci.* **2015**, *6*, 966. [CrossRef] [PubMed]

137. Mulekar, J.J.; Huq, E. Expanding roles of protein kinase CK2 in regulating plant growth and development. *J. Exp. Bot.* **2014**, *65*, 2883–2893. [CrossRef] [PubMed]

138. Montenarh, M.; Götz, C. Ecto-protein kinase CK2, the neglected form of CK2 (Review). *Biomed. Rep.* **2018**, *8*, 307–313. [CrossRef] [PubMed]

139. Gosti, F.; Beaudoin, N.; Serizet, C.; Webb, A.A.; Vartanian, N.; Giraudat, J. ABI1 protein phosphatase 2C is a negative regulator of abscisic acid signaling. *Plant Cell* **1999**, *11*, 1897–1910. [CrossRef] [PubMed]

140. Ludwików, A. Targeting proteins for proteasomal degradation—A new function of *Arabidopsis* ABI1 protein phosphatase 2C. *Front. Plant Sci.* **2015**, *6*, 310. [CrossRef] [PubMed]

141. Chibani, K.; Ali-Rachedi, S.; Job, C.; Job, D.; Jullien, M.J.; Grappin, P. Proteomic analysis of seed dormancy in *Arabidopsis*. *Plant Physiol.* **2006**, *142*, 1493–1510. [CrossRef] [PubMed]

142. Gallardo, K.; Job, C.; Groot, S.P.; Puype, M.; Demol, H.; Vandekerckhove, J.; Job, D. Proteomics of *Arabidopsis* seed germination. A comparative study of wild-type and gibberellin-deficient seeds. *Plant Physiol.* **2002**, *129*, 823–837. [CrossRef] [PubMed]

143. Lehesranta, S.J.; Davies, H.V.; Shepherd, L.V.T.; Koistinen, K.M.; Massat, N.; Nunan, N.; McNicol, J.W.; Kärenlampi, S.O. Proteomic analysis of the potato tuber life cycle. *Proteomics* **2006**, *6*, 6042–6052. [CrossRef] [PubMed]

144. Bachem, C.; Van der Hoeven, R.; Lucker, J.; Oomen, R.; Casarini, E.; Jacobsen, E.; Visser, R. Functional genomic analysis of potato tuber life-cycle. *Potato Res.* **2000**, *43*, 297–312. [CrossRef]

145. Ronning, C.M.; Stegalkina, S.S.; Ascenzi, R.A.; Bougri, O.; Hart, A.L.; Utterbach, T.R.; Vanaken, S.E.; Riedmuller, S.B.; White, J.A.; Cho, J.; et al. Comparative analyses of potato expressed sequence tag libraries. *Plant Physiol.* **2003**, *131*, 419–429. [CrossRef] [PubMed]

146. Campbell, M.; Segear, E.; Beers, L.; Knauber, D.; Suttle, J. Dormancy in potato tuber meristems: Chemically induced cessation in dormancy matches the natural process based on transcript profiles. *Funct. Integr. Genom.* **2008**, *8*, 317–328. [CrossRef] [PubMed]

147. Baginsky, S. Plant proteomics: Concepts, applications, and novel strategies for data interpretation. *Mass Spectrom. Rev.* **2009**, *28*, 93–120. [CrossRef] [PubMed]

148. Strasser, R. Biological significance of complex N-glycans in plants and their impact on plant physiology. *Front. Plant Sci.* **2014**, *5*, 363. [CrossRef] [PubMed]

149. Strasser, R. Plant protein glycosylation. *Glycobiology* **2016**, *26*, 926–939. [CrossRef] [PubMed]

150. Lerouge, P.; Cabanes-Macheteau, M.; Rayon, C.; Fischette-Laine, A.C.; Gomord, V.; Faye, L. N-glycoprotein biosynthesis in plants: Recent developments and future trends. *Plant Mol. Biol.* **1998**, *38*, 31–48. [CrossRef] [PubMed]

151. Koshiyama, I. Carbohydrate component in 7S protein of soybean casein fraction. *Agric. Biol. Chem.* **1966**, *30*, 646–650. [CrossRef]

152. Ericson, M.C.; Chrispeels, M.J. Isolation and characterization of glucosamine-containing storage glycoproteins from the cotyledons of *Phaseolus aureus*. *Plant Physiol.* **1973**, *52*, 98–104. [CrossRef] [PubMed]

153. Basha, S.M.M.; Beevers, L. Glycoprotein metabolism in the cotyledons of *Pisum sativum* during development and germination. *Plant. Physiol.* **1976**, *57*, 93–97. [CrossRef] [PubMed]

154. Hall, T.C.; Mcleester, R.C.; Bliss, F.A. Equal expression of the maternal and paternal alleles for the polypeptide subunits of the major storage protein of the bean *Phaseolus vulgaris* L. *Plant Physiol.* **1977**, *59*, 1122–1124. [CrossRef] [PubMed]

155. Eaton-Mordas, C.A.; Moore, K.G. Seed glycoproteins of *Lupinus angustifolius*. *Phytochemistry* **1978**, *17*, 619–621. [CrossRef]

156. Badenoch-Jones, J.; Spencer, D.; Higgins, T.J.V.; Millerd, A. The role of glycosylation in storage-proteins synthesis in developing pea seeds. *Planta* **1981**, *153*, 201–209. [CrossRef] [PubMed]

157. Sengupta, C.; Deluca, V.; Bailey, D.S.; Verma, D.P.S. Post-translational processing of 7S and 11S components of soybean storage proteins. *Plant Mol. Biol.* **1981**, *1*, 19–34. [CrossRef] [PubMed]

158. Weber, E.; Manteuffel, R.; Jakubek, M.; Neumann, D. Comparative studies on protein bodies and storage proteins of *Pisum sativum* L. and *Vicia faba* L. *Biochem. Physiol. Pflanzen* **1981**, *176*, 342–356. [CrossRef]

159. Chrispeels, M.J.; Higgins, T.J.V.; Craig, S.; Spencer, D. Role of the endoplasmic reticulum in the synthesis of reserve proteins and the kinetics of their transport to protein bodies in developing pea cotyledons. *J. Cell Biol.* **1982**, *93*, 5–14. [CrossRef] [PubMed]

160. Bollini, R.; Vitale, A.; Chrispeels, M.J. In vivo and in vitro processing of seed reserve protein in the endoplasmic reticulum: Evidence for two glycosylation steps. *J. Cell Biol.* **1983**, *96*, 999–1007. [CrossRef] [PubMed]

161. Lioi, L.; Bollini, R. Contribution of processing events to the molecular heterogeneity of four banding types of phaseolin, the major storage protein of *Phaseolus vulgaris* L. *Plant Mol. Biol.* **1984**, *3*, 345–353. [CrossRef] [PubMed]

162. Paaren, H.E.; Slightom, J.L.; Hall, T.C.; Inglis, A.S.; Blagrove, R.J. Purification of a seed glycoprotein: N-terminal and deglycosylation analysis of phaseolin. *Phytochemistry* **1987**, *26*, 335–343. [CrossRef]

163. Sturm, A.; Van Kuik, J.A.; Vliegenthart, J.F.G.; Chrispeels, M.J. Structure, position, and biosynthesis of the high mannose and complex oligosaccharide chains of the bean storage protein phaseolin. *J. Biol. Chem.* **1987**, *262*, 13392–13403. [PubMed]

164. Duranti, M.; Guerrieri, N.; Takajashi, T.; Cerletti, P. The legumin-like storage proteins of *Lupinus albus* seeds. *Phytochemistry* **1988**, *27*, 15–23. [CrossRef]

165. Duranti, M.; Gorinstein, S.; Cerletti, P. Rapid separation and detection of concanavalin. A reacting glycoproteins: Application to storage proteins of a legume seed. *J. Food Biochem.* **1990**, *14*, 327–330. [CrossRef]

166. Lawrence, M.C.; Suzuki, E.; Varghese, J.N.; Davis, P.C.; Van Donkelaar, A.; Tulloch, P.A.; Colman, P.M. The three-dimensional structure of the seed storage protein phaseolin at 3 Å resolution. *EMBO J.* **1990**, *9*, 9–15. [CrossRef] [PubMed]

167. Duranti, M.; Guerrieri, N.; Cerletti, P.; Vecchio, G. The legumin precursor from white lupin seed. *Eur. J. Biochem.* **1992**, *206*, 941–947. [CrossRef] [PubMed]

168. Duranti, M.; Gius, C.; Sessa, F.; Vecchio, G. The saccharide chain of lupin seed conglutin γ is not responsible for the protection of the native protein from degradation by trypsin, but facilitates the refolding of the acid-treated protein to the resistant conformation. *Eur. J. Biochem.* **1995**, *230*, 886–891. [CrossRef] [PubMed]

169. Duranti, M.; Horstmann, C.; Gilroy, J.; Croy, R.R.D. The molecular basis for N-glycosylation in the 11S globulin (legumin) of lupin seed. *J. Protein Chem.* **1995**, *14*, 107–110. [CrossRef] [PubMed]

170. Kolarich, D.; Altmann, F. N-glycan analysis by matrix-assisted laser desorption/ionization mass spectrometry of electrophoretically separated nonmammalian proteins: Application to peanut allergen Ara h 1 and olive pollen allergen Ole e 1. *Anal. Biochem.* **2000**, *285*, 64–75. [CrossRef] [PubMed]

171. López-Torrejón, G.; Salcedo, G.; Martín-Esteban, M.; Díaz-Perales, A.; Pascual, C.Y.; Sánchez-Monge, R. Len c 1, a major allergen and vicilin from lentil seeds: Protein isolation and cDNA cloning. *J. Allergy Clin. Immunol.* **2003**, *112*, 1208–1215. [CrossRef] [PubMed]

172. Lauer, I.; Foetisch, K.; Kolarich, D.; Ballmer-Weber, B.K.; Conti, A.; Altmann, F.; Vieths, S.; Scheurer, S. Hazelnut (*Corylus avellana*) vicilin Cor a 11: Molecular characterization of a glycoprotein and its allergenic activity. *Biochem. J.* **2004**, *383*, 327–334. [CrossRef] [PubMed]

173. Vaz, A.C.; Pinheiro, C.; Martins, J.M.N.; Ricardo, C.P.P. Cultivar discrimination of Portuguese *Lupinus albus* by seed protein electrophoresis: The importance of considering "glutelins" and glycoproteins. *Field Crop Res.* **2004**, *87*, 23–34. [CrossRef]

174. Fukuda, T.; Prak, K.; Fujioka, M.; Maruyama, N.; Utsumi, S. Physicochemical properties of native adzuki bean (*Vigna angularis*) 7S globulin and the molecular cloning of its cDNA isoforms. *J. Agric. Food Chem.* **2007**, *55*, 3667–3674. [CrossRef] [PubMed]

175. Marsh, J.T.; Tryfona, T.; Powers, S.J.; Stephens, E.; Dupree, P.; Shewry, P.R.; Lovegrove, A. Determination of the N-glycosylation patterns of seed proteins: Applications to determine the authenticity and substantial equivalence of genetically modified (GM) crops. *J. Agric. Food Chem.* **2011**, *59*, 8779–8788. [CrossRef] [PubMed]

176. Picariello, G.; Amigo-Benavent, M.; del Castillo, M.D.; Ferranti, P. Structural characterization of the N-glycosylation of individual soybean β-conglycinin subunits. *J. Chromatogr. A* **2013**, *1313*, 96–102. [CrossRef] [PubMed]

177. Santos, C.N.; Alves, M.; Oliveira, A.; Ferreira, R.B. β-N-acetylhexosaminidase involvement in α-conglutin mobilization in *Lupinus albus*. *J. Plant Physiol.* **2013**, *170*, 1047–1056. [CrossRef] [PubMed]

178. Schiarea, S.; Arnoldi, L.; Fanelli, R.; Combarieu, E.; Chiabrando, C. In-depth glycoproteomic characterization of γ-conglutin by high-resolution accurate mass spectrometry. *PLoS ONE* **2013**, *8*, e73906. [CrossRef] [PubMed]

179. Kumari, N.; Kofi, K.J.; Grimbs, S.; D'Souza, R.N.; Kuhnert, N.; Vrancken, G.; Ullrich, M.S. Biochemical fate of vicilin storage protein during fermentation and drying of cocoa beans. *Food Res. Int.* **2016**, *90*, 53–65. [CrossRef] [PubMed]

180. Snégaroff, J.; Bouchez, I.; Smaali, M.I.A.; Pecquet, C.; Raison-Peyron, N.; Jolivet, P.; Laurière, M. Barley γ3-hordein: Glycosylation at an atypical site, disulfide bridge analysis, and reactivity with IgE from patients allergic to wheat. *Biochim. Biophys. Acta* **2013**, *1834*, 395–403. [CrossRef] [PubMed]

181. Sathe, S.K.; Hamaker, B.R.; Sze-Tao, K.W.C.; Venkatachalam, M. Isolation, purification, and biochemical characterization of a novel water soluble protein from inca peanut (*Plukenetia volubilis* L.). *J. Agric. Food Chem.* **2002**, *50*, 4906–4908. [CrossRef] [PubMed]

182. Kishimoto, T.; Watanabe, M.; Mitsui, T.; Mori, H. Glutelin basic subunits have a mammalian mucin type O-linked disaccharide side chain. *Arch. Biochem. Biophys.* **1999**, *370*, 271–277. [CrossRef] [PubMed]

183. Racusen, D.; Foote, M.A. A major soluble glycoprotein from potato tubers. *J. Food Biochem.* **1980**, *4*, 43–52. [CrossRef]

184. Bauw, G.; Nielsen, H.V.; Emmersen, J.; Nielsen, K.L.; Jørgensen, M.; Welinder, K.G. Patatin, Kunitz protease inhibitors and other major proteins in tuber of potato cv. Kuras. *FEBS J.* **2006**, *273*, 3569–3584. [CrossRef] [PubMed]

185. Welinder, K.G.; Jørgensen, M. Covalent structures of potato tuber lipases (patatins) and implications for vacuolar important. *J. Biol. Chem.* **2009**, *284*, 9764–9769. [CrossRef] [PubMed]

186. Lattová, E.; Brabcová, A.; Bártová, V.; Potěšil, D.; Bárta, J.; Zdráhal, Z. N-glycome profiling of patatins from different potato species of *Solanum* genus. *J. Agric. Food Chem.* **2015**, *63*, 3243–3250. [CrossRef] [PubMed]

187. Allen, L.W.; Svenson, R.H.; Yachnin, S. Purification of mitogenic proteins derived from *Phaseolus vulgaris*: Isolation of potent and weak phytohemagglutinins possessing mitogenic activity. *Proc. Natl. Acad. Sci. USA* **1969**, *63*, 334–341. [CrossRef] [PubMed]

188. Miller, J.B.; Hsu, R.; Heinrikson, R.; Yachnin, S. Extensive homology between the subunits of the phytohemagglutinin mitogenic proteins derived from *Phaseolus vulgaris*. *Proc. Natl. Acad. Sci. USA* **1975**, *72*, 1388–1391. [CrossRef] [PubMed]

189. Vitale, A.; Chrispeels, M.J. Transient N-acetylglucosamine in the biosynthesis of phytohemagglutinin: Attachment in the Golgi apparatus and removal in protein bodies. *J. Cell Biol.* **1984**, *99*, 133–140. [CrossRef] [PubMed]

190. Faye, L.; Sturm, A.; Bollini, R.; Vitale, A.; Chrispeels, M.J. The position of the oligosaccharide side-chains of phytohemagglutinin and their accessibility to glycosidases determines their subsequent processing in the Golgi. *Eur. J. Biochem.* **1986**, *158*, 655–661. [CrossRef] [PubMed]

191. Sturm, A.; Chrispeels, M.J. The high mannose oligosaccharide of phytohemagglutinin is attached to asparagine 12 and the modified oligosaccharide to asparagine 60. *Plant Physiol.* **1986**, *80*, 320–322. [CrossRef]

192. Sturm, A.; Bergwerff, A.A.; Vliegenthart, J.F.G. H-NMR structural determination of the N-linked carbohydrate chains on glycopeptides obtained from the bean lectin phytohemagglutinin. *Eur. J. Biochem.* **1992**, *204*, 313–316. [CrossRef] [PubMed]

193. Sharma, M.; Vishwanathreddy, H.; Sindhura, B.R.; Kamalanathan, A.S.; Swamy, B.M.; Inamdar, S.R. Purification, characterization and biological significance of mannose binding lectin from *Dioscorea bulbifera* bulbils. *Int. J. Biol. Macromol.* **2017**, *102*, 1146–1155. [CrossRef] [PubMed]

194. Waglay, A.; Karboune, S.; Alli, I. Potato protein isolates: Recovery and characterization of their properties. *Food Chem.* **2014**, *142*, 373–382. [CrossRef] [PubMed]

195. Zarkadas, C.G.; Gagnon, C.; Gleddie, S.; Khanizadeh, S.; Cober, E.R.; Guillemette, R.J.D. Assessment of the protein quality of fourteen soybean [*Glycine max* (L.) Merr.] cultivars using amino acid analysis and two-dimensional electrophoresis. *Food Res. Int.* **2007**, *40*, 129–146. [CrossRef]

196. Kirihara, J.A.; Hunsperger, J.P.; Mahoney, W.C.; Messing, J.W. Differential expression of a gene for a methionine-rich storage protein in maize. *Mol. Gen. Genet.* **1988**, *211*, 477–484. [CrossRef] [PubMed]

197. Gibbon, B.C.; Wang, X.; Larkins, B.A. Altered starch structure is associated with endosperm modification in Quality Protein Maize. *Proc. Natl. Acad. Sci. USA* **2003**, *100*, 15329–15334. [CrossRef] [PubMed]

198. El-Shemy, H.A.; Khalafalla, M.M.; Fujita, K.; Ishimoto, M. Improvement of protein quality in transgenic soybean plants. *Biol. Plant.* **2007**, *51*, 277–284. [CrossRef]

199. Jiang, C.; Cheng, Z.; Zhang, C.; Yu, T.; Zhong, Q.; Shen, J.; Huang, X. Proteomic analysis of seed storage proteins in wild rice species of the *Oryza genus*. *Proteome Sci.* **2014**, *12*, 51. [CrossRef] [PubMed]

200. Jiang, M.X.; Zhai, L.J.; Yang, H.; Zhai, S.M.; Zhai, C.K. Analysis of active components and proteomics of chinese wild rice (*Zizania latifolia* (Griseb) *Turcz*) and *Indica* rice (*Nagina22*). *J. Med. Food.* **2016**, *19*, 798–804. [CrossRef] [PubMed]

201. Hedrick, P.H. *Genetics of Populations*, 3rd ed.; Jones and Bartlett: Sudbury, MA, USA, 2005; pp. 525–595. ISBN 0-7637-4772-6.

202. Bárta, J.; Bártová, V. Patatin, the major protein of potato (*Solanum tuberosum* L.) tubers, and its occurrence as genotype effect: Processing versus table potatoes. *Czech J. Food Sci.* **2008**, *26*, 347–359. [CrossRef]

203. Bártová, V.; Bárta, J.; Brabcová, A.; Zdráhal, Z.; Horáčková, V. Amino acid composition and nutritional value of four cultivated South American potato species. *J. Food Compos. Anal.* **2015**, *40*, 78–85. [CrossRef]

204. Ogawa, T.; Tayama, E.; Kitamura, K.; Kaizuma, N. Genetic improvement of seed storage proteins using three variant alleles of 7S globulin subunits in soybean (*Glycine max* L.). *Jpn. J. Breed.* **1989**, *39*, 137–147. [CrossRef]

205. Gobbetti, M.; Giuseppe Rizzello, C.; Di Cagno, R.; De Angelis, M. Sourdough lactobacilli and celiac disease. *Food Microbiol.* **2007**, *24*, 187–196. [CrossRef] [PubMed]

206. Tanner, G.J.; Blundell, M.J.; Colgrave, M.L.; Howitt, C.A. Creation of the first ultra-low gluten barley (*Hordeum vulgare* L.) for coeliac and gluten-intolerant populations. *Plant Biotechnol. J.* **2016**, *14*, 1139–1150. [CrossRef]

207. Wild, D.; Robins, G.G.; Burley, V.J.; Howdle, P.D. Evidence of high sugar intake, and low fibre and mineral intake, in the gluten-free diet. *Aliment. Pharmacol. Ther.* **2010**, *32*, 573–581. [CrossRef] [PubMed]

208. Öhlund, K.; Olsson, C.; Hernell, O.; Öhlund, I. Dietary shortcomings in children on a gluten-free diet. *J. Hum. Nutr. Diet.* **2010**, *23*, 294–300. [CrossRef] [PubMed]

209. Kawaura, K.; Miura, M.; Kamei, Y.; Ikeda, T.M.; Ogihara, Y. Molecular characterization of gliadins of Chinese Spring wheat in relation to celiac disease elicitors. *Genes Genet. Syst.* **2018**. [CrossRef] [PubMed]

210. van den Broeck, H.C.; Gilissen, L.J.W.J.; Smulders, M.J.M.; van der Meer, I.M.; Hamer, R.J. Dough quality of bread wheat lacking α-gliadins with celiac disease epitopes and addition of celiac-safe avenins to improve dough quality. *J. Cereal Sci.* **2011**, *53*, 206–216. [CrossRef]

211. Rizzello, C.G.; Curiel, J.A.; Nionelli, L.; Vincentini, O.; Di Cagno, R.; Silano, M.; Gobbetti, M.; Coda, R. Use of fungal proteases and selected sourdough lactic acid bacteria for making wheat bread with an intermediate content of gluten. *Food Microbiol.* **2014**, *37*, 59–68. [CrossRef] [PubMed]

212. Bewley, J.D.; Black, M. *Seeds: Physiology of Development and Germination*, 2nd ed.; Plenum Press: New York, NY, USA, 1994; pp. 377–416. ISBN 978-1-4899-1002-8.

213. Sugliani, M.; Rajjou, L.; Clerkx, E.J.M.; Koornneef, M.; Soppe, W.J.J. Natural modifiers of seed longevity in the *Arabidopsis* mutants abscisic acid insensitive3-5(*abi3-5*) and leafy cotyledon1-3(*lec1-3*). *New Phytol.* **2009**, *184*, 898–908. [CrossRef] [PubMed]

214. Dobiesz, M.; Piotrowicz-Cieślak, A.I.; Michalczyk, D.J. Physiological and biochemical parameters of lupin seed subjected to 29 years of storage. *Crop Sci.* **2017**, *57*, 2149–2159. [CrossRef]

215. Rajjou, L.; Lovigny, Y.; Groot, S.P.C.; Belghazi, M.; Job, C.; Job, D. Proteome-wide characterization of seed aging in *Arabidopsis*: A comparison between artificial and natural aging protocols. *Plant Physiol.* **2008**, *148*, 620–641. [CrossRef] [PubMed]

216. Sano, N.; Rajjou, L.; North, H.M.; Debeaujon, I.; Marion-Poll, A.; Seo, M. Staying alive: Molecular aspects of seed longevity. *Plant Cell Physiol.* **2015**, *57*, 660–674. [CrossRef] [PubMed]

217. Kalemba, E.M.; Pukacka, S. Carbonylated proteins accumulated as vitality decreases during long-term storage of beech (*Fagus sylvatica* L.) seeds. *Trees* **2014**, *28*, 503–515. [CrossRef]

218. Senakoon, W.; Nuchadomrong, S.; Chiou, R.Y.Y.; Senawong, G.; Jogloy, S.; Songsri, P.; Patanothai, A. Identification of peanut seed prolamins with an antifungal role by 2D-GE and drought treatment. *Biosci. Biotechnol. Biochem.* **2015**, *79*, 1771–1778. [CrossRef] [PubMed]

219. Hajheidari, M.; Eivazi, A.; Buchanan, B.B.; Wong, J.H.; Majidi, I.; Salekdeh, G.H. Proteomics uncovers a role for redox in drought tolerance in wheat. *J. Proteome Res.* **2007**, *6*, 1451–1460. [CrossRef] [PubMed]

220. Yahata, E.; Maruyama-Funatsuki, W.; Nishio, Z.; Tabiki, T.; Takata, K.; Yamamoto, Y.; Tanida, M.; Saruyama, H. Wheat cultivar-specific proteins in grain revealed by 2-DE and their application to cultivar identification of flour. *Proteomics* **2005**, *5*, 3942–3953. [CrossRef] [PubMed]

221. Kim, Y.S.; Lee, Y.H.; Kim, H.S.; Kim, M.S.; Hahn, K.W.; Ko, J.H.; Joung, H.; Jeon, J.H. Development of patatin knockdown potato tubers using RNA interference (RNAi) technology, for the production of human-therapeutic glycoproteins. *BMC Biotechnol.* **2008**, *8*, 36. [CrossRef] [PubMed]

MDPI

St. Alban-Anlage 66

4052 Basel

Switzerland

Tel. +41 61 683 77 34

Fax +41 61 302 89 18

www.mdpi.com

Molecules Editorial Office

E-mail: molecules@mdpi.com

www.mdpi.com/journal/molecules